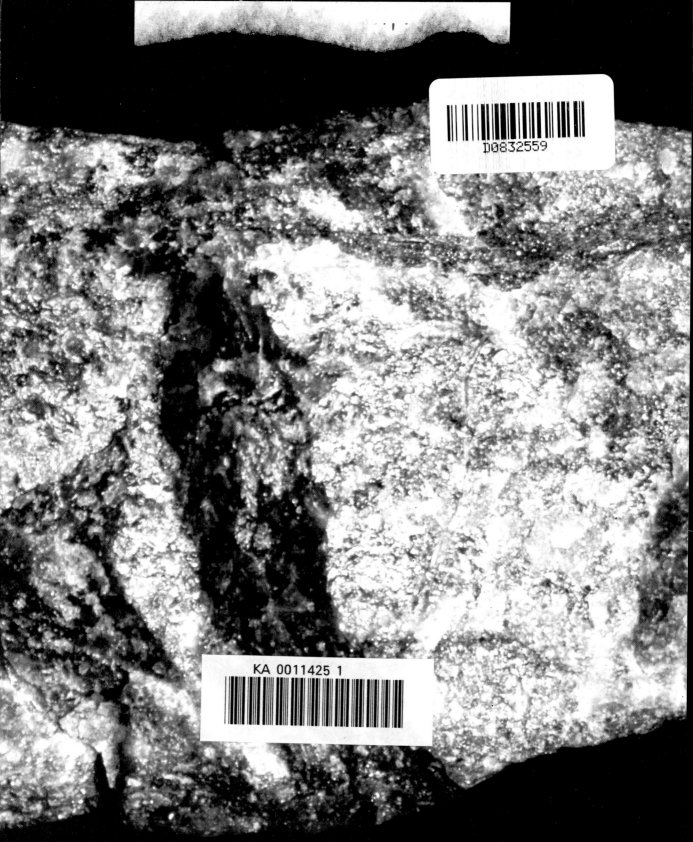

CONSOLIDATED GOLD FIELDS

A CENTENARY PORTRAIT

WEIDENFELD AND NICOLSON LONDON

Paul Johnson

CONSOLIDATED
Gold Fields

A CENTENARY
PORTRAIT

To Simon and Virginia

CONSOLIDATED GOLD FIELDS

Principal Subsidiary Companies

ARC (100%)

ARC America Corporation (100%)

Gold Fields Mining Corporation (100%)

Principal Associated Companies

Renison Goldfields Consolidated (49%)

Gold Fields of South Africa (48%)

Newmont Mining Corporation (26%)

TITLE PAGE A blast at Ortiz in New Mexico, the first gold mine developed by GFMC in the United States.

Copyright © 1987 Consolidated Gold Fields PLC

Published in Great Britain by
George Weidenfeld & Nicolson Limited
91 Clapham High Street
London SW4 7TA

ISBN 0 297 78967 8

Phototypeset by Keyspools Limited, Golborne, Lancs.

Printed and bound in Italy by L.E.G.O. Vicenza

Contents

Foreword by Rudolph Agnew 7

Introduction: Capitalism and Creativity 8

1. The Rise and Development of Consolidated Gold Fields 13

2. South Africa: Achievements, Problems and Opportunities 54

3. Gold Fields' Operations in America 104

4. The Newmont Investment 137

5. Building for the Future in Australia 158

6. ARC Limited: A British Success Story 201

7. The Central Strategy of Consolidated Gold Fields 228

8. Moving into the Twenty-First Century 251

Acknowledgements 253

Index 254

Foreword

As a part of our celebration of the centenary of Consolidated Gold Fields I asked Paul Johnson to sketch the people and businesses which now make up the complex, international confederation which is the direct descendant of the bold, entrepreneurial investment into the new, raw gold fields of South Africa made one hundred years ago.

This is Paul Johnson's independent impressionist portrait, not an exhausting catalogue of mines and investments nor a pious, comprehensive chronicle. The skeleton is there with enough history and current detail to add depth and substance to the image. I believe that the author's snapshots give us a lively, colourful view of the complexity, contradictions, tensions and problems inherent in a large, international group today.

This book will help everyone to understand more clearly what Gold Fields is today and where it is going as it starts on its second century.

Rudolph I. J. Agnew

Introduction: Capitalism and Creativity

One of the strongest and most precious of human impulses is the need to create. Some men and women express it through painting and sculpture, by writing books or composing music. To others the opportunity is given to create wealth: to enrich humanity by taking the crude materials of which the earth is composed and transforming them into useful commodities and objects. In the establishment and growth of a business, the need to create is probably a more powerful motive than the desire to make money. In this sense capital has a moral status because it is the means whereby the creative process can proceed. Capitalism has a moral authority because it is the best way so far discovered to make the process efficient. In turn, a business fulfils a moral purpose not only by creating wealth itself but by enabling those it employs to share in the activity, to exercise their wits and skills, their energies and ideas, in improving the quality of the products and services of the firm, and so its fortunes. It is a central characteristic of a well-run business that it gives its own people the satisfaction of creation.

Consolidated Gold Fields is a worldwide group engaged in an exceptionally wide range of creative activities. Together with its subsidiaries and associates it is a highly complex financial institution designed to find and exploit natural resources and turn them into wealth. It provides the money so that these operations can employ the latest technology and the highest human skills, and it reinvests a substantial part of its profits in discovering new resources and turning them into fresh ventures.

In 1984–1985, I travelled 20,000 miles examining the properties of the group. It employs nearly 100,000 people to produce hundreds of different products, spread over six continents and working all the time zones of the world. At midnight, Greenwich Mean Time, ARC Marine's dredger *Deepstone*, back from the East Anglian Coast with 7,500 tons of aggregate, is heading for Southampton, ready to discharge its cargo at the rate of 2,000 tons an hour. At Eneabba, north of Perth in Western Australia, the morning shift is reporting for duty in the synthetic rutile plant. Thousands of feet below the surface in the Transvaal, at the great West Driefontein gold mine, teams of miners working at more than sixty faces are clearing up after the afternoon's blasting. In Clearwater, Florida, computer operators at the Central Dispatch department of Cement Products Corporation, are checking evening maintenance schedules for sixty giant ready-mixed concrete transporters. In Tasmania, at the Renison Tin Mine, the morning shift is just coming to the surface. The Group's geologists are setting up their equipment in the California desert, getting ready for lunch in Papua New Guinea, settling down for the night in the Orange Free State. CGF is indeed a commonwealth on which the sun never sets.

These travels have left certain especially vivid images on my mind. There was a conversation with the Assistant Manager of the Black Mountain Mine in Northern Cape Province. It produces lead, copper and zinc, with silver in the concentrates. We had just visited the vast cavern at the seventh level, watching a multiple-drill machine boring shot-holes in the hanging wall, another at work on the side wall, while a scoop-truck was lifting five tons of ore at a time and dropping it into a carrier. Each of these enormous vehicles had to manoeuvre with delicate precision in the confined space, so that we beheld a brutish quadrille of mechanical mastodons. The noise was indescribable. But down below on the eighth level where they were building new underground work shops, we were able to talk.

The Assistant Manager told me: 'Having a workshop underground will make a huge difference to costs. Those machines you saw take a terrific hammering. They need extensive daily maintenance and often repairs. Each time we take one up to the surface it consumes twenty-seven litres of fuel. That's quite apart from delays and congestions on the incline as you shift the vehicles up and down.' He reeled off, one by one, half a dozen other ways they had found to shave production costs. He recited the facts and figures, the percentages and savings, with the intensity of a Benedictine monk intoning the liturgy. As I watched his face I became aware, perhaps for the first time, that the need to make a success of a mine could inspire the same kind of dedication – one might almost call it an obsession – which a painter acquires when he tackles a really ambitious canvas. Cutting costs, the use of brainpower and original ideas to make the productive process more efficient, is a form of creativity too.

In a house in the suburbs of Johannesburg, I talked for hours with a Consulting Engineer who has designed and supervised the creation of one of the most successful gold-mines in the world. He had worked all his life to make deep-level gold-mines more cost-effective, but above all less prodigal in human life and muscle-power. He had many triumphs. But he had one grievance. He told me: 'In a deep-level gold-mine the depth to which the main shaft can be sunk is limited because, beyond a certain point, the steel winding ropes will fail under their own weight. This limit is in excess of 8,000 feet but it's better to stop at about 6,000 feet in order to be able to hoist an effective payload. This means you must sink a sub-shaft and to do this you excavate an underground winding engine chamber and bring all your shaft-sinking equipment underground. That's expensive. And time-consuming. I had the idea of using high-tensile steel ropes for sinking the adjacent ventilation shaft to well beyond the required depth with a view to utilising the lower portion as a sub-shaft. In effect this would mean that the sub-shaft could be sunk from surface whilst sinking of the large main shaft was still in progress. This would save at least two years in reaching the planned depth of the mine.

'However, we had a horizontal connection to an adjoining group mine from which development was being carried out in order to block out ore reserves. This was being done so that the mine could be brought to production immediately on completion of the main shaft. Unfortunately water problems associated with bad ground delayed the sinking of the main shaft and it soon became obvious that the mine would not be producing revenue on due date as planned. Difficulty was experienced in raising additional funds so the powers-that-be decided that the sinking of the sub-shaft should be stopped forthwith and that the ventilation shaft be used for hoisting ore as a temporary expedient – albeit on a smaller scale than had been planned for the main shaft. This decision constituted the biggest technical disappointment of my career. It really hurt to see a good technical scheme sacrificed on the altar of financial necessity. A short-term cash flow problem was overcome but, due to the time required to complete the first sub-shaft, years elapsed before the mine could be brought to full production.'

Just outside the new town of Milton Keynes in Buckinghamshire, ARC is creating a wildfowl shelter and much else from a group of worked-out quarries. I was shown round it by the Warden, a big man, of over six feet tall, with a bushy beard. He was a farmer by training, a naturalist by taste: a creator, too. He now has ten lakes which are gradually being transformed for different purposes: fishing, sailing, speed-boats, and some for the wildfowl. What was once a lunar landscape of industrial ruin is gradually assuming the contours of an ordered landscape. In time it will be a place of domesticated beauty. The Warden runs the farm, which when I was there had nearly 300 sheep, bred for spring lambs and to keep the parkland trim. I have seldom met a man more suited to the job he was doing, or more content with it. He lives in a little stone house overlooking one of his lakes and, standing there with me, surveying his domain, he said simply: 'I must be the happiest man in England.'

There is another memory from Enid, Oklahoma, at the George E. Failing Company, where they make vibrator trucks and other exploration equipment. It is not Gold Fields' policy to remain in manufacturing, the company has now been sold and rightly so, but it can seldom have owned a more impressive property than this medium-sized firm, 'the Cadillac of the rig business'. The President asked me to sit in on a conference call to China, to settle the final details for clinching the big export order. He told me: 'This is a critical call to our man out there. We have to get the figures right. We approach customers from a cost viewpoint. It is especially vital when dealing with the Chinese, who are the best traders in the world. You have to sense how far you can go without losing your pants.' The President and four executives sat around his desk, the speaker switched on, notebooks and calculators ready. There were two other links by phone to offices in the building, and the line to China. The call came through. The calculating and decision-making began. There was tension. Then, as the complexities began to resolve themselves, jokes. The President noted: 'These Chinese guys, they're going all bureaucratical. That's because they're getting capitalist, like us.' I looked at the intent faces round the desk. They were taking part in a scene which marked the climax of months of detailed preparation: careful costing and design changes, marketing, salesmanship. Now everything was being put to the test and the result was in sight. They were enjoying it! It was like watching a group of professional musicians, after many rehearsals, giving the first performance of a new work.

Again, I recall an underground classroom in a colliery at Coalbrook in the

Orange Free State. A black instructor was teaching eight black miners the first principles of loss-control. Coalbrook is a mine with a long shadow: in the early 1960s, long before Gold Fields owned it, a cataclysm killed more than 400 miners there. No one who goes back to those days now works at the pit, but the need for safety is graven on all hearts. I listened to the instructor. He was very methodical, but eager and vehement too. Loss-control is a much more sophisticated concept than old-fashioned safety drill. It puts avoidable risk and danger in its true perspective as one of many forms of inefficiency. It shows how everyone, by performing their specific roles precisely, can help to avoid not just loss of life and limb but loss of time, material and the product. Increased safety thus becomes an automatic by-product of an operation where everyone is thoroughly trained and responsible; it is a function of an intelligent and educated workforce, and of an efficient management. Here were eight young men, who had come straight from their villages, being introduced to one of the subtlest notions of advanced industrial society. They were rapt by these new ideas. The instructor was absorbed in punching his message home. For a few moments I forgot that our little academy was surrounded by coal and rock, hundreds of feet below the surface.

These are some of the images that remain with me. They all concern people. A business is primarily about people. It must be capitalised and organised and directed to unleash to the maximum extent the creative energies of the people it employs. At Gold Fields' headquarters, the company strategists deploy its financial resources to achieve this end. If they get their calculations right, and give their people the opportunities to create, the products and the profits will follow. Here then is a portrait of Consolidated Gold Fields: a portrait of the company itself and the Gold Fields Group, its origins, history and present objectives; and a portrait of the multitude of men and women who are helping to achieve them.

Cecil John Rhodes (1853–1902), the co-founder of Consolidated Gold Fields

1 The Rise and Development of Consolidated Gold Fields

CECIL JOHN RHODES, the co-founder of Consolidated Gold Fields, was the archetype of the businessman-creator. Karl Marx, in Chapter 23 of *Das Kapital*, defines the capitalist as an automaton under the control of his own money: 'his actions are a mere function of the capital which, through his instrumentality, is endowed with will and consciousness, so that his own private consumption must be regarded by him as a robbery perpetrated on accumulation.' Rhodes was not like that at all. He was born in 1853, son of a Hertfordshire vicar, and was sent, for health reasons, to South Africa in 1870, aged seventeen, to work on his brother's cotton farm. He seems to have been born with an overwhelming desire to create huge enterprises, and to impose order on chaos. To him, the acquisition of money was merely a means to these ends. His values were not materialistic at all. He had a life-long passion for education, to him part of the ordering process. When he first went prospecting, he took with him a pick, two spades, six volumes of the classics and a Greek lexicon. During the period when he was founding his business and acquiring his first solid capital, he constantly interrupted his work by going to Oxford to keep terms and take a degree. When he died, he left his entire fortune of £6 million to the public service, most of it to found colleges, universities and scholarships.

Rhodes was a huge man, well over six feet tall, with an enormously broad and deep chest, massive head and spectacular brow, often compared to Napoleon's. People were overwhelmed by him. No one was less of an orator; but when he spoke, men were transfixed. His letters ignore punctuation, grammar, syntax; but he had an extraordinary talent for the striking phrase. Above all, he had a genius for conciliation and negotiation; he could bring rivals and antagonists together as though by mesmerism – it was part of his creative gift.

He came to Africa at the perfect time: just ahead of the rush. The 'Dark Continent', as Sir Henry Stanley called it, was opening up. The year after Rhodes arrived, Stanley 'found' Livingstone, a world sensation at Ujiji. His books, essays in brilliant journalism, enthralled a huge public, not least potential investors: *How I Found Livingstone* (1872), *Through the Dark Continent* (1878), *In Darkest Africa* (1890). World demand for tropical products, especially coffee, tea, sugar and rubber, was rising fast. The great powers were alerted, then rapidly became rapacious. Starting in 1883 and over a mere eighteen months, Germany annexed four vast areas in South-West and East Africa: Belgium swallowed up an empire in the Congo. At the Berlin Conference in 1884 Britain acquiesced in these depredations in return for German support of her occupation of Egypt and in anticipation of her own expansion in the south and east. Everywhere, agents of the powers

ABOVE Part of a letter from Rhodes, written in 1895 to H.E.M.Davies, the first secretary of Gold Fields and later chairman of the company. The letter deals with his purchase of shares in the Cape Town Waterworks Company which gave him ownership of a large area on the slopes of Table Mountain and Devil's Peak, where he built his house. The house, Groote Schuur, was left to the people of South Africa and became the official residence of South African Prime Ministers.

ABOVE Rhodes at Groote Schuur.

RIGHT An 'account rendered' to Rhodes on 17 August 1886 for supplies purchased during the first prospecting trip to the new Transvaal gold find. As the letter heading shows, the diggers' camp had already been named 'Johannesburg'. It consisted of little more than a few wood-and-iron shacks. A case of whiskey, it will be seen, cost £3.15s. in 1886.

were buying 'treaties' from native sovereigns.

When Rhodes arrived in 1870, South Africa was an agricultural backwater. This phase ended with a bang the same year. Diamonds had been found three years before, but the discoveries in 1870 at Vooruitzich Farm (a name soon changed to Kimberley, after the Colonial Secretary, who found the original one unpronounceable) were much more substantial. Rhodes's brother, Herbert, hurried there in January 1871, to stake a claim; Rhodes himself followed in October. The next year he formed a partnership with the owner of an adjoining claim, Charles Rudd. Rudd was nine years Rhodes's senior, the son of Henry Rudd, a shipbuilder, of Hanworth Hall in Norfolk. Their meeting was a fortunate conjunction. Rudd was more experienced than Rhodes, and had what Rhodes felt he conspicuously lacked, a good education (Rudd had been at Harrow and Trinity, Cambridge). Rhodes was the merchant-adventurer writ large, the vision-

ABOVE Charles Durrell Rudd, co-founder of Consolidated Gold Fields, at his desk.

LEFT Charles Rudd (1844–1916), nine years Rhodes' senior and the son of a Norfolk shipbuilder, who met Rhodes when they owned adjoining claims at Kimberley. They became partners in 1872.

ary, the colossal risk-taker. Rudd was the calm and cautious man of business, the master of the painstaking detail, a superb chairman of company meetings.

Diamonds were of critical importance both to Rhodes and to the development of South Africa. Rhodes loved diamonds. He always wrote the word with a capital 'D'. He never had the same feeling for gold. Diamonds, at any rate in the first stage, could be dug out cheaply and sold quickly. Thus cash flowed into South Africa for the first time in quantity. In Cape Town, the Government Secretary slammed a diamond onto the table of the House of Assembly and declared: 'Gentlemen, this is the rock on which the future success of South Africa will be built.' He was right. By 1872, South Africa was exporting diamonds worth £1,600,000 a year. Griqualand, hitherto inhabited only by a few hundred Griqua under their chief Waterboer, was swamped by 45,000 whites and three times as many blacks. Kimberley opened up a stock exchange and the City of London began trading in South African shares.

The diamond boom was important too because it foreshadowed the mining finance house system and introduced organisational characteristics which Consolidated Gold Fields bears to this day. Hitherto, throughout the world, major discoveries in precious stones and metals had produced wasteful boom-bust cycles which scratched at the surface of the deposits and often left nothing substantial behind. The Kimberley boom threatened to be exactly the same. By 1875 there were 3,600 separate claims in Kimberley, shared by more than 20,000 individual owners. Men just dug deeper into enormous holes in the ground, hauling the earth up on thousands of separate ropes and pulleys. Photographs of the two big holes at Kimberley present an un-forgettable image of primeval chaos and blind greed. It revolted Rhodes to his very soul. He said soon after he arrived: 'The time is coming when the small man will have to go. These pits cannot be worked much deeper. We shall have to mine the ground on the largest possible scale.'

It was Rhodes's great merit that he grasped this point long before anyone else, and had the will-power and the skill to make large-scale diamond mining a reality. He specialised in pumps for deep working; his partner Rudd in wire-ropes. In 1875 the diggings first went below the 300-foot mark, and that brought the first shake-out of small claim-owners. Men sold their claims; so did the banks; and Rhodes and Rudd bought. It took Rhodes ten years of hard negotiating and capital-raising to rationalise the diamond mining industry, but at the end of it he had reduced the claims from 3,600 to 100 and created De Beers Consolidated Mines, which included his greatest rival, Barney Barnato. This huge firm was well on its way to regulating the worldwide sale of diamonds (as it does to this day). It was capable of raising capital for the very largest operations, and not just in the field of diamonds. Rhodes himself had already established a reputation in the City of London as a man who could be relied upon to make money for investors, as well as himself (by 1880 his annual income was over £50,000).

The establishment of a rationalised diamond industry with access to the London capital market was perfectly timed to coincide with the first major gold discoveries in South Africa. Diamonds were the detonator which made

OPPOSITE ABOVE Charles Rudd in Kimberley in 1874 with his wife Frances (seated) and child.

OPPOSITE BELOW Charles Rudd and his family outside their home in Kimberley in 1874. At that time he was looking after Rhodes' interests at the diamond diggings while Rhodes was studying at Oriel College, Oxford. His university experience led him to establish the famous Rhodes scholarships.

Rhodes created De Beers Consolidated Mines, the world's largest diamond producer after acquiring the interests of his greatest rival, Barney Barnato (below).

ABOVE AND OPPOSITE
By 1875 the diamond boom
was at its height. There were
3,600 separate claims in
Kimberley shared by more
than 20,000 owners. Men dug
deeper and deeper holes,
hauling up earth on thousands
of ropes and pulleys. Rhodes
soon realised that the 'small'
man had to go because mining
needed to be on the largest
possible scale.

possible the real industrial-mining explosion, based on gold. To give an idea
of the comparative financial scale, the first century of South African
diamond-mining, 1867–1967, recovered stones worth £700 million; the first
eighty-two years on the Rand alone, 1886–1978, yielded gold worth £6,000
million.

Gold was not new to Africa. The Egyptians had got African gold in the
second millenium BC. Africans worked in base as well as precious metals.
The Sotho, for instance, practised smelting as early as AD 800. They and
other tribes had a metal folklore. They identified the process of smelting

KIMBERLEY-MINE, 1874.

with the reproductive cycle of women, who were not allowed near a forge lest they bewitch the iron. Among the tribes, the Venda were the specialists in gold. Everyone knew there was gold in southern Africa. Gold was worked on the quartz reefs of the Transvaal in 1853. Alluvial gold was found on the Olifants River in 1868, at Marasbad in Zoutpansberg in 1873, in the eastern Transvaal at Pilgrim's Rest in 1872. All these produced minor rushes; but such patchy, limited and unsystematically worked discoveries, mainly of alluvial gold, raised expectations they could not satisfy. There was another rush in 1882–84 in the eastern Transvaal to a place soon dignified by the name of a new town, Barberton. Its Bray's Golden Quarry yielded 50,000 ounces of gold from 13,000 tons of ore, a record; the nearby Sheba Mine was said to be the richest on earth. But only a handful of the Barberton mines ever paid; the town, stock exchange and all, went bust.

Hence when, on 5 June 1885, Harry Struben, speaking on the tennis court behind the Union Club in Pretoria, first told an assemblage of Transvaal dignitaries and businessmen about the gold deposits of the Witwatersrand, the reaction was sceptical. Yet the Witwatersrand is quite different from any other gold field in the world. It poses mining problems which were unique in the 1880s and are formidable even today. The reefs slope deeply below the surface in very complex formations. The gold is uniform in composition but it is microscopically fine and often mixed up with pyrites – then an insoluble

Cecil Rhodes and his brothers Herbert and Frank at the diamond diggings at Colesburg Kopje, Cape Colony, in 1872.
Left to right: J.E.Dick-Lauder, N.Garstin, H.MacLeod, Dr W.G.Atherstone, Herbert Rhodes, Frank Rhodes, Cecil Rhodes, and R.A.Nesbitt.

metallurgical problem. What is more, the average gold content per ton is very low indeed. On the other hand, the field is enormous, by far the largest in the world: 170 miles by 100 miles. Recovering its riches meant mining on the largest possible scale and employing hitherto undreamed-of sums of capital. If sufficient capital could be effectively invested, the rewards would be correspondingly enormous.

The diamond houses of Kimberley were the obvious centres from which to finance the mining of the Rand for gold. But in the light of the recent Barberton débâcle there was little enthusiasm to begin with. Rhodes's enemy, J.B.Robinson, was the first to invest heavily. He borrowed £20,000 from a Rhodes' associate, Alfred Beit, and proved exceptionally lucky. At a time when an ounce of gold recovered per ton of ore was regarded as payable, he got ten ounces a ton from his mine. Rhodes and Rudd moved to the Rand in 1886 and began buying Boer farms. Rhodes prided himself on his skill in persuading these dour men to part with their acres at reasonable prices. Later in life, he told many stories about his diplomatic triumphs. He clinched one deal, he said, by suddenly throwing in as a major concession six wagon-loads of firewood. Another he sealed by offering the farmer a glass of French brandy, only to be told: '*Ik gebruik nie brandewyn maar ik zal blievers een blikje jem neem*' ('I never drink brandy but I'll take a tin of jam instead.') Rhodes thus acquired seven farms in likely areas, but he had no geologists to guide him.

With this porfolio of properties, Rudd was sent to London in November 1886 to set up a company and raise capital. Gold Fields of South Africa Limited, the original name of Consolidated Gold Fields, was registered on 9 February 1887 and its first statutory meeting held the following month. All 100,000 shares were promptly subscribed. There were two reasons for this. Rhodes and Rudd were not greedy, and they were seen not to be greedy. They risked their own money as well as the shareholders', and in a financial epoch so savagely described in Anthony Trollope's *The Way We Live Now*, that was an important point. Whereas other companies formed in 1887 to exploit the Rand insisted that the bulk of the capital subscribed be declared 'vendors' interest', which meant that promoters got from fifty to seventy-five per cent of all the proceeds of flotation, Rhodes and Rudd handed over the properties they had acquired to the company, receiving in return not cash, all of which was used as working capital, but Founders' Shares. It is true they then got three-fifteenths of any profit made, which in time brought them huge sums; but until the company began to make a profit they got nothing at all. The second reason was more subtle. To London investors, the name Rhodes meant money. It was a magic word, and it retained its power until his death. Quite why this should have been so is impossible to say. Rhodes never met his shareholders or attended a company meeting. He despised them: 'I have no intention of working for these fellows for the balance of my life. A more ungrateful crew I have never come across.' But they seem to have sensed, as though by instinct, the creative spirit in this extraordinary man, and backed it with their cash. Rhodes's ability to conjure money out of the London pavements, merely on the strength of his name, became his biggest single asset.

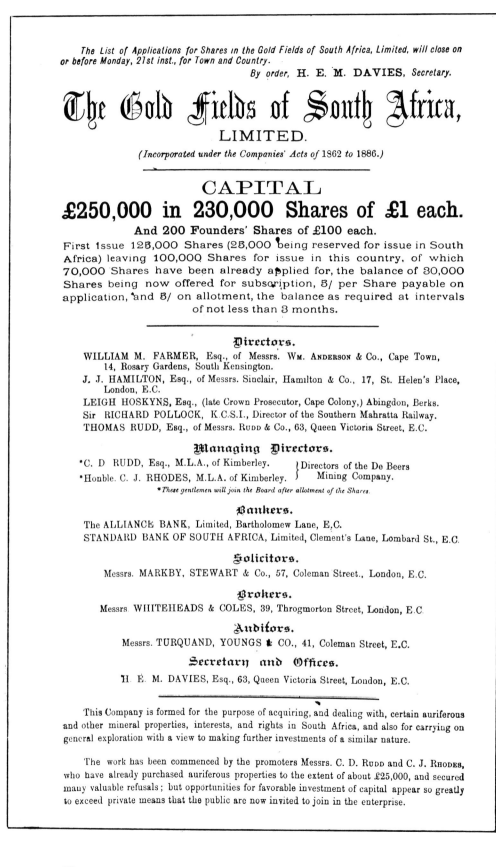

The List of Applications for Shares in the Gold Fields of South Africa, Limited, will close on or before Monday, 21st inst., for Town and Country.

By order, H. E. M. DAVIES, *Secretary.*

The Gold Fields of South Africa,
LIMITED.

(Incorporated under the Companies' Acts of 1862 to 1886.)

CAPITAL
£250,000 in 230,000 Shares of £1 each.
And 200 Founders' Shares of £100 each.

First issue 125,000 Shares (25,000 being reserved for issue in South Africa) leaving 100,000 Shares for issue in this country, of which 70,000 Shares have been already applied for, the balance of 30,000 Shares being now offered for subscription, 5/ per Share payable on application, and 5/ on allotment, the balance as required at intervals of not less than 3 months.

Directors.

WILLIAM M. FARMER, Esq., of Messrs. WM. ANDERSON & Co., Cape Town, 14, Rosary Gardens, South Kensington.

J. J. HAMILTON, Esq., of Messrs. Sinclair, Hamilton & Co., 17, St. Helen's Place, London, E.C.

LEIGH HOSKYNS, Esq., (late Crown Prosecutor, Cape Colony,) Abingdon, Berks.

Sir RICHARD POLLOCK, K.C.S.I., Director of the Southern Mahratta Railway.

THOMAS RUDD, Esq., of Messrs. RUDD & Co., 63, Queen Victoria Street, E.C.

Managing Directors.

*C. D RUDD, Esq., M.L.A., of Kimberley.　　} Directors of the De Beers

*Honble. C. J. RHODES, M.L.A. of Kimberley. } 　Mining Company.

* These gentlemen will join the Board after allotment of the Shares.

Bankers.

The ALLIANCE BANK, Limited, Bartholomew Lane, E.C.

STANDARD BANK OF SOUTH AFRICA, Limited, Clement's Lane, Lombard St., E.C.

Solicitors.

Messrs. MARKBY, STEWART & Co., 57, Coleman Street., London, E.C.

Brokers.

Messrs. WHITEHEADS & COLES, 39, Throgmorton Street, London, E.C.

Auditors.

Messrs. TURQUAND, YOUNGS & CO., 41, Coleman Street, E.C.

Secretary and Offices.

H. E. M. DAVIES, Esq., 63, Queen Victoria Street, London, E.C.

This Company is formed for the purpose of acquiring, and dealing with, certain auriferous and other mineral properties, interests, and rights in South Africa, and also for carrying on general exploration with a view to making further investments of a similar nature.

The work has been commenced by the promoters Messrs. C. D. RUDD and C. J. RHODES, who have already purchased auriferous properties to the extent of about £25,000, and secured many valuable refusals; but opportunities for favorable investment of capital appear so greatly to exceed private means that the public are now invited to join in the enterprise.

That was just as well, for the assets he had bought, on which Gold Fields was floated, proved not to be gold fields at all. Nor is this surprising because buying land in the areas was a blind, hit-or-miss affair. The geology of the field was not understood at all. The miners of 1885–86 thought the gold-bearing outcrops they found to be a form of river gravel. Albert Truter, formerly of Gold Fields, one of South Africa's leading geologists, explained to me: 'The gold of the Witwatersrand occurs in a layered pile of sedimentary rocks approximately 25,000 feet (about 5 miles) thick, occupying an elongated basin some 200 miles long and 100 miles wide that formed in the earth's crust some 2,700 million years ago. The rocks were derived from sediments (mud, sand and gravel) that were brought from surrounding higher ground into the basin by flowing water and other mechanical agencies. The basin was filled with water, so there was wave action which spread the immigrant sediments in successive layers of mud, sand and gravel over the gradually subsiding basin floor, each layer being repeated many times. In time these sediments were consolidated and indurated into rocks; the mud becoming shale, the sand quartzite and the gravel conglomerate.

'The gold was derived from the same source as the other sediments and was transported into the basin simultaneously with and by the same means as the others. During the transportation process the gold, being heavy, got mixed with the other less easily moved material, namely the gravel, and has remained there throughout all subsequent vicissitudes of earth history to become a fossil constituent of the now famous conglomerate layers.

'The conglomerate layers are characteristically thin, individual thicknesses ranging from less than an inch to merely a few feet, and they are sandwiched between much thicker layers of quartzite which, like the shale, often measure several hundreds, even thousands, of feet.

'Most of the conglomerates contain only a little gold and are not of economic value. Some contain no gold at all, but a few are very rich. These occur chiefly along the northern and western margin of the basin, but they persist for great distances (several miles) towards the central parts of the basin. They have supported the incomparable Witwatersrand gold-mining industry for a hundred years and enormous untouched reserves still remain in the ground.

'The Witwatersrand rocks became buried under tremendous thicknesses (several miles in total) of younger rocks. These include further shales and quartzites, igneous lavas and dolomite, which sometimes contains extraordinarily large amounts of ground water constituting a serious mining hazard. Some of this cover of younger rocks was subsequently removed through erosion so that in a few places the Witwatersrand rocks are again exposed at surface.

'This applies particularly along the northern rim of the basin where the gold-bearing conglomerate reefs outcrop along an eighty miles long crescent of low hills, the Witwatersrand (white watershed ridge), on which the city of Johannesburg and several other satellite cities and towns were built during the last century. It was on these outcrops that gold was discovered and the first mining began.'

OPPOSITE The front page of the Gold Fields of South Africa's original prospectus issued in 1887. Following a reorganisation, this company became Consolidated Gold Fields of South Africa in August 1982.

But of course the outcrop gold was only a tiny fraction of the whole: the reefs sloped down to great depths towards the centre of the basin, and recovering the bulk of the gold meant mining at depths man had never hitherto attempted. If the discoveries had come twenty years before, they could not have been exploited. But a remarkable conjunction of events and technologies enabled the Rand to produce the very large-scale, continuous and even production which effectively confirmed gold, rather than silver, as the world currency standard. The first factor was the availability of Kimberley diamond money, with the backing of the City of London. The second was cheap Natal coal, which in the 1880s came on-stream in large quantities. The third was the opening up of the African interior from about 1885 onwards, which made available the plentiful labour without which deep-level mining was not possible: the number of Africans in the mines rose from 14,000 in 1890 to nearly 100,000 in 1899. The fourth was the development of the machinery essential to deep-level, large-scale mining: the pumps and cables, above all the big stamp-mills. It was only in late 1887, when the first big stamp-mill began processing the ore, that men began to grasp the true potential of the Rand. By the end of 1888 there were 688 stamps running, crushing 15,000 tons of ore a month and producing 200,000 ounces of gold a year. In the first three years of using the new machinery, 1887–89, the Rand produced 642,803 ounces of gold. By 1894 there were over 2,000 stamps crushing 2,750,000 tons annually and producing gold

ABOVE One of the first gold prospectors' camps.

OPPOSITE ABOVE The first gold 'battery' used to crush gold-bearing rocks in the Transvaal.

OPPOSITE BELOW A day's shooting near Pretoria.

worth £7 million. A year later, the capitalised value of Rand gold mines was £71 million, rising to £82.5 million at the height of the 1890s boom. The big companies were reckoned to be worth over £100 million, and the Rand was now producing twenty per cent, rising fast, of the world's gold supply.

The way in which Gold Fields of South Africa eventually surfaced in a favourable position in this maelstrom of activity did not reflect well on Rhodes's judgement. Rudd was forced to admit at the annual meeting of November 1889 that they had failed to secure any rich gold-bearing properties. Their first crushing results had been disastrous, a mere ten dwts (pennyweights) to the ton. Gold Fields, it is true, earned a profit of £23,503 in its first full year of trading; but this was obtained by creating gold-mining

THE RECENT DISCOVERIES OF GOLD IN THE TRANSVAAL

The first gold miners pictures in *The Graphic*, 26 February 1887.

companies from portions of its estates, and selling off interests in them. After they got the poor crushing results, they invested most of the company's cash in diamonds, which produced good profits but annoyed the shareholders, who had the gold lust. The first Rand gold boom peaked in spring 1889, when workings on the Main Reef hit pyritic ore, which meant building chlorination plants and adding £1 a ton to the cost of processing the ore. The boom bust, many of the 400 smaller companies went to the wall, banks with them. Rhodes had been warned in good time by Beit that a bust was coming. He instructed the company's secretary H.E.M.Davies, who acted as its financial manager, to sell out of gold shares before the market collapsed. The money was put into sound diamond shares. Hence Gold Fields was saved financially but by 1890 virtually all its money was in diamonds and it did not control a single gold mine.

By 1891–92 geological surveys had given plain indications of the size and continuity of the Main Reef at depth, and in the following year it was reached by a borehole at 2,343 feet below ground. At roughly the same time, the Macarthur-Forrest cyanide process became available, enabling low-grade pyritic ore to be processed cheaply. In the first half of the decade, the East Rand coal field was developed, and rail outlets driven through from the Cape (1892) and Delagoa Bay on the east coast (1895). Gold Fields' mining engineer, Percy Tarbutt, was now convinced that deep-level mining was

Sketch 8 shows the mouth of the Incline Shaft, Robinson Gold Mining Company. These drawings appeared in the *Illustrated London News*, 6 July 1889.

GFSA rebuilt its head office in Fox Street, Johannesburg, during the early 1980s.

both technically and economically feasible, but initially he found Rhodes a difficult man to convince. Tarbutt himself and two partners created the African Estates Agency and were the first to peg deep-level claims. They were followed by Beit, who pegged 1,357 claims, which eventually became Rand Mines; and his initial investment of £200,000 swiftly rose in value to over £50 million.

It was Beit who finally persuaded Rhodes to invest in deep-level mining. He wanted Rhodes in his schemes because no one else had such ready and unfailing access to London capital. So when Rand Mines, the first big deep-level company, was formed in 1893, Gold Fields got ten per cent, plus ten per cent of Founders' Profits, one of the best investments it ever made. Rhodes now decided the time had come to rationalise the Rand gold field as he had earlier straightened out diamond mining.

He set Davies, as the financial expert, to work. What Davies did was to create a new company, which consisted of Gold Fields of South Africa, Tarbutt's African Estates Agency, the African Gold Investment Company and the South African Gold Trust and Agency Company. The amalgamation was called Consolidated Gold Fields of South Africa Limited, with a capital of £1,250,000. Rhodes and Rudd agreed to have their Founders' Shares bought out for £80,000 in CGFSA shares plus a further 25,000 shares sold at par for cash.

This then was the first real mining finance house. Financially it was an amazing success. In its first full year it made a profit of £207,455, transferring £50,000 to reserves and paying out ten per cent to shareholders. Profits rose to £308,963 in the second year, and the company was able to raise £600,000 in debentures to sink deep-level shafts. In 1895 it made a stupendous profit of £2,161,778, paying a final dividend of 100 per cent (making 125 per cent for the year), carrying forward £1,145,000 and putting £200,000 to reserve. Hitherto, no public limited liability company in the world had made a profit of £2 million.

Rhodes afterwards said that whereas it had taken him twenty years to rationalise Kimberley diamond mining, the Rand was sorted out in a mere five. Certainly, the example of CGFSA was rapidly followed. The following year, the House of Eckstein became Rand Mines Limited. In 1895 G. and L. Albu became General Mining. In 1905 the House of Wernher Beit became Central Mining. Finally, in 1918, Messrs Ad Goerz & Co became Union Corporation. The firm called Johannesburg Consolidated Investments (known as 'Johnnies') went through a similar process of amalgamation without changing its name.

What were the characteristics of a mining finance house? It was the parent of the group, holding a controlling interest in quite separate mining, exploration and prospecting companies. It did not itself search for or produce gold. What it did was to provide for the financial needs of its dependent companies. It acted as their banker. It raised capital for them on the market. It also provided for their administrative needs and supplied high-grade technical services for the group. It had two boards, one in London mainly concerned with money, and one in Johannesburg mainly concerned with mining.

The first office of Gold Fields of South Africa in Johannesburg. It was built in 1888 in what was to become the fashionable residential suburb of Doornfontein.

Deep-level mining on the Rand was by its nature very costly and risky. The mining finance system dealt with both these problems by making possible a low-cost structure and by spreading the risk as widely as was practicable. A mining finance group might have an interest in as many as fifteen or sixteen mines. Moreover, the groups could and did diversify into other areas and industries, thus spreading the risk still further. Ernest Oppenheimer, founder of the Anglo-American group, noted in 1930 the value of the mining finance house system. On the Rand, he said, where 'there are many companies whose properties adjoin or are adjacent to one another, all engaged in the same class of work, the existence of a central organisation for the supply of expert advice and information on matters of common interest is clearly of incalculable value. It certainly ensures the companies great economies.'

The value of the system of groups was enhanced, and the general interest protected, by a series of powerful communal bodies. The Witwatersrand Chamber of Mines (1889) negotiated with government, on the one hand, and saw to the supply of labour, on the other, through the Witwatersrand Native Labour Association and the Native Recruiting Corporation Limited, which trawled over a large part of southern Africa for potential African miners. The sign over its Zululand depot read: *Abathanda imali, abathanda izinkhomonidhela elula eye eGoli; nanti iHovisi* – 'Lovers of money, lovers of cattle, the road is easy to go to the City of Gold; here is the office'. The function of the Chamber of Mines and its related bodies was not merely to ensure the supply of labour but to reduce its cost, which was forty to forty-five per cent of total production costs by 1895, by eliminating rival bidding. It ensured that uniform wage rates were negotiated at a high level. The point was made bluntly by F.H.Hatch and J.A.Chambers in their authoritative survey *The Gold Mines of the Rand* (1895), when they wrote of 'the probability that wages paid to natives will be eventually reduced, through the unremitting efforts of the Chamber of Mines which, since its inception, has directed much energy to the question of the *reduction* of native wages and the *increase* in the supply of native labour.'

The virtues of the entire system of groups and Chamber were summed up by another expert, Henry Clay, thus: 'The gold-mining industry may claim to provide a working model of a "rationalised" industry. Through the group system of control of separate mining companies and the close co-operation of the whole industry through the Chamber of Mines and its subsidiary services, it has substituted for the blind selection by competition of the fittest to survive, a conscious and deliberate choice of methods, equipment, areas and personnel on the basis of an extremely detailed comparative study of results.'

Rhodes was prominent in this reduction of chaos to profitable order. But the creation of the mining finance house system was only part of his schemes for southern Africa; indeed, its chief function in his eyes was to supply the finance-power for much bigger ventures. It was characteristic of him that all the major companies he founded had the widest possible aims. Their articles of association empowered them to engage in virtually any form of activity, including the acquisition of territory. The truth is that, once diamonds were discovered in quantity at Kimberley, and still more so after the opening up of the Rand gold-field, big business in South Africa and geopolitics were inseparable.

Britain had promptly annexed the diamond fields of Griqualand in 1871, and in 1880 the territory was incorporated in Cape Colony, Rhodes being elected to one of the two new parliamentary seats thereby created, a seat he held till his death. Three years before, in his first will dated 19 September 1877, he had set forth his world programme, bequeathing his property to:

> the establishment, promotion and development of a Secret Society the aim and object whereof shall be the extension of British rule throughout the world, the perfecting of a system of emigration from the United Kingdom and of colonisation by British subjects of all lands where the means of livelihood are attainable by energy, labour and enterprise, and especially the occupation by British settlers of the entire continent of Africa, the Holy Land, the Valley of the Euphrates, the islands of Cyprus and Candia, the whole of South America, the islands of the Pacific not heretofore possessed by Great Britain, the whole of the Malay Archipelago, the seaboard of China and Japan, the ultimate recovery of the United States of America as an integral part of the British Empire, the inauguration of a system of colonial representation in the Imperial parliament, which may tend to weld together the disjointed members of the empire, and finally the foundation of so great a power as hereafter to render wars impossible and promote the best interests of humanity.

This twenty-four-year-old's vision was soon succeeded by a more limited and practical one of a union of southern Africa under Britain. He told the Cape Parliament: 'We should be the dominant state in South Africa and should carry out the union of the South African states.' In 1883–84 he was the driving force in Cape Colony expansion and annexation to the north, first in Basutoland, then Bechuanaland. The discovery that southern Africa was mineral-rich inevitably adumbrated the dispossession or break-up of all the major black monarchies and a shift towards open white rule everywhere. The discovery of massive gold reserves in the Transvaal meant another geopolitical shift, away from Britain towards the Boer republics. From being the poorest of the South African states, the Transvaal suddenly became the richest – its revenue increased twenty-five times, 1883–95 – and most aggressive. What Rhodes feared most of all was a predatory alliance between the Boer states and the rapidly expanding German colonies, which would become the dominant military power in the sub-continent, halt the British advance northwards, and eventually drive the British into a mere Cape enclave.

Not that Rhodes was anti-Boer. He built up his political power in Cape Colony through an alliance between the British and the Cape Dutch. 'The Dutch are the coming race in South Africa', he said, 'and must share in running the country.... You cannot have real prosperity ... until you have first established complete confidence between the two races.' But Britain must be the controlling power, not least for the sake of the blacks, whose fate would be desperate under German-Boer paramountcy. Rhodes felt that it was right the blacks should be introduced to industrial society. He told the Cape parliament in 1894: 'It is the duty of the government to remove these poor children out of their state of sloth and laziness and give them some gentle stimulants to go forth and find out something of the dignity of labour.' His notion was to educate the blacks, both technically and in self-government. He became Prime Minister of the Cape in 1890 and introduced into the native territories village and district councils, levying rates to set up schools. His 1892 Franchise Act defined eligible voters as those who could read and write and were in receipt of a labourer's wages, irrespective of colour.

There were two ways in which Rhodes could exorcise the German-Boer nightmare and he tried each of them in turn. The first was to leapfrog over the Boer states and establish British rule to the north. As he put it in 1888, immediately after the founding of Gold Fields: Here are the politics of South Africa in a nutshell. We must adopt the whole responsibility for the interior ... the gist of the South African question lies in the extension of the Cape Colony to the Zambesi.' This territory of Matabeleland was a tyrannical black empire, ruled with great cruelty by Lobengula. The Reverend John Moffat, British deputy-commissioner and the most experienced missionary in those parts, despaired of the fate of the helot tribes, like the Ndebele, in the king's possessions: 'It will be a blessing to the world when they are broken up.' Encouraged by Rhodes, Moffat negotiated with Lobengula a treaty (11 February 1888) under which he undertook not to enter into agreements with other powers or cede anything without the permission of the High Commissioner at the Cape. Rhodes followed this by dispatching Rudd to the king's kraal in October – one step ahead of emissaries from Germany and the Transvaal – and obtained a commercial treaty which granted him and his associates sole rights to the minerals throughout the king's possessions. In return the king got £1,200 a year, 1,000 rifles and 100,000 rounds of ammunition, Rhodes throwing in a steamboat, *gratis* (it was never actually delivered).

As Rhodes put it, 'Our concession is so gigantic it is like giving a man the

OPPOSITE ABOVE At Gold
Fields headquarters in
Johannesburg, Rhodes and
leaders of the mining industry
(the Reform Committee)
planned the British takeover of
the Transvaal from the Boers.
Dr Jameson was sent from
Rhodesia in 1895 with a
column of yeomen-cavalry to
encourage an uprising in the
Transvaal. The uprising never
took place and Jameson
surrendered two days after
crossing the Transvaal border.

On the day Dr Jameson
surrendered Bettington's
Horse gathered outside Gold
Fields Offices in Johannesburg
before setting out to ride to
Jameson's rescue. They were
quickly recalled. Subsequently,
both Rhodes and Rudd had to
resign as joint Managing
Directors of Consolidated
Gold Fields.

OPPOSITE BELOW The scene
in the court room in Pretoria
during the trial of the Reform
Committee which had planned
the raid.

whole of Australia.' Rhodes wanted to turn the territory into a British protectorate, but the government in London would not agree with him. But they did agree to his alternative suggestion that he set up a chartered company, on the lines of the old East India Company, and on 29 October 1889 the British South Africa Company was granted a royal charter. From the start, Rhodes associated Gold Fields with the venture. It was given the biggest share of the original concession, eight-and-a-half portions out of thirty; and when the 'Chartered Company', as it was known, was incorporated, Gold Fields got 25,500 shares. The creation of this great enterprise won universal applause from the British. Moffat said he thought 'it is God who is working out what has come about in a way that the will of no man could have accomplished.' *The Times* saw it as a supreme example of combining in the name of progress 'British capitalism and philanthropy'. To the *Pall Mall Gazette* it was 'civilisation by company'. Rhodes put it more succinctly and brutally: 'Philanthropy plus Five Per cent'. In 1890 Rhodes sent in a 900-strong pioneer column, under one of his gold-mining associates, Dr Leander Starr Jameson, which effectively established a military presence. Three years later, the company's forces intervened in a tribal war between the Mashonas and the Matabele. Lobengula was dispossessed and died (Rhodes brought three of Lobengula's sons to the Cape to be educated at his expense) and the company took over the rule of 440,000 square miles, the nucleus of the future Rhodesia.

Having thus sealed off German or Boer expansion to the north, Rhodes prepared the second blow: the takover of the Transvaal. British capital was creating a new Transvaal, symbolised by its vast new city of gold, Johannesburg. British subjects provided more than ninety per cent of the state's wealth: but they were *Uitlanders*, denied the vote; the constitution and suffrage were specifically designed to exclude them from power. The Transvaal government was irredeemably corrupt and inefficient. Its President, Paul Kruger, was everything that Rhodes detested: illiberal, obscurantist, anti-progress. He was a founder of the extreme fundamentalist sect, the Dopper Kerk, and believed to his dying day that the earth was flat.

By 1895, the peak of the first really big gold boom, the *Uitlanders* felt their position had become intolerable. They used their Reform Committee, pledged to secure a more liberal franchise, to prepare an internal uprising. It was to a great extent managed from Gold Fields' headquarters in Johannesburg (a quarter of a century later, in the 1920s, cases of rifles were discovered in the cellars). Rhodes determined to give the projected rising external assistance. He authorised Dr Jameson, who had managed for him the destruction of Lobengula, to collect a column of yeoman-cavalry, and he drew into the net Alfred Beit's company, Rand Mines, and its chairman, Lionel Phillips. Among the leading plotters were Colonel Frank Rhodes, Cecil's brother, and the famous American mining engineer, John Hays Hammond, whom Rhodes had hired as a consultant for Gold Fields at the princely salary of £15,000 a year. There was plenty of brainpower behind the *putsch* but not enough simple military discipline. Jameson's column of 500 crossed into the Transvaal on 29 December 1895 and surrendered two

ABOVE Leander Starr Jameson, leader of the Jameson raid.

Betting on the Horse. — Leaving to meet Jameson

1914

days later. The internal uprising never took place. The mismanagement of the raid was one of the great tragedies in the history of Africa. Had it succeeded, and the Transvaal passed under British rule, the Boer War would never have taken place, the Union would have come about without a struggle, and South Africa would have pursued a course of development into an independent multi-racial society much closer to the normal pattern of the British Commonwealth. *Apartheid* and its evil fruits would have remained a figment in the imagination of a few Boer intellectuals.

But as President Kennedy was to remark, 'Victory has a thousand fathers; defeat is an orphan.' The year 1895 was Rhodes's *annus mirabilis*: he was joint managing Director of Consolidated Gold Fields of South Africa, the most profitable public company in the world; Managing Director of the Chartered Company and Northern Railways; Chairman of De Beers, and Prime Minister of Cape Colony. Had the Raid succeeded, it would have been hailed as a stroke of genius, on a level with his Matabele concession. But it failed, and everyone turned on him. Rudd, who had not been in on the plot, reported to Gold Fields it 'could only have been arranged by a madman'. Jameson served time in Wormwood Scrubbs. Frank Rhodes, Hammond and Phillips were sentenced to death in Pretoria, but let off with fines of £25,000 each.

OPPOSITE ABOVE An ore train at Simmer and Jack, named after August Simmer and John Jack. It was John Jack who took ore from this mine to Glasgow University where the cyanide method of gold extraction was discovered.

OPPOSITE BELOW Robinson Deep – the key producer in South Johannesburg – helped Gold Fields to prosper in the immediate post-Rhodes era.

Robinson Deep, with Sub-Nigel (left) and Simmer and Jack were known as Faith, Hope and Charity because they were the mainstay of Gold Fields for 25 years at the turn of the century.

THE MORNING MARKET JOHANNESBURG.

Johannesburg in 1890 –
already a thronging
metropolis.

It was the end of the Rhodes-Rudd era at Gold Fields. Both men resigned as joint Managing Directors. They also had to renounce their right to two-fifteenths of the profits, though both remained on the board. At an extraordinary general meeting of the company on 31 March 1896, the capital was increased by 100,000 new shares to £1,975,000. In return for their forfeiture of special rights, Rhodes and Rudd got 100,000 shares, worth between £1.2 million and £1.5 million. The operation was skilfully managed by Davies, who also solved the problem of the chairmanship. It had been occupied by Rudd's brother, Thomas, who was now anxious to retire as part of the shakeout. Davies found Lord Harris, a former Governor of Bombay, who became Vice-Chairman in July 1896 and Chairman three years later. This marked the point at which Gold Fields ceased to be heavily and directly involved in politics, put behind its fast-moving buccaneering origins, and became a respectable public company like any other.

Or did it? The truth is a mining finance house, then as now, cannot afford to renounce the spirit of adventure. It must take risks, often huge risks. Mines are born to die. New ones must be found and developed in good time. The most successful mining usually takes place near the frontiers of civilisation, or beyond them, and is peculiarly vulnerable to change and chance and political accident. The Boer War, which the Jameson Raid might have averted, opened in October 1899 with the Boer invasion of Natal. All the Rand mines shut down. Johannesburg fell to the British on 31 August 1900, but the guerrilla phase of the conflict lasted a further eighteen months, and the mines did not reopen until March 1902. During this long hiatus, not only was no gold being produced but the mining companies had to pay and equip 1,500 mine guards. Gold Fields paid a handsome dividend in 1898; nothing at all in 1899–1901, and it again passed a dividend in 1903, as post-war depression, labour shortages and the unwillingness of the London capital market to have anything to do with South Africa led to the postponement of ambitious schemes.

In the immediate post-Rhodes era, the company's prosperity was due mainly to the advice of John Hays Hammond. His claim that he pioneered deep-level mining on the Rand is not true; but he certainly accelerated the development of vast, deep-level mines and involved Gold Fields in the most advanced and ambitious projects. He enlisted Gold Fields, together with two other houses, Wernher, Beit and S. Neumann, in a company called Gold Fields Deep Limited, which eventually controlled all the mining areas of the central Rand and in particular the key producer in south Johannesburg, Robinson Deep. He also took the company into the area east of Johannesburg, by staking out deep-level claims and then, in 1894, merging them with an existing mine named after its owners, August Simmer, a Bavarian, and John Jack, a Scot. The Simmer and Jack mine was rapidly and successfully developed by Gold Fields and became one of the world's greatest gold-mines, with a life of seventy years and, in its day, the most consistent major profit-maker on the Rand. Further east still, some thirty

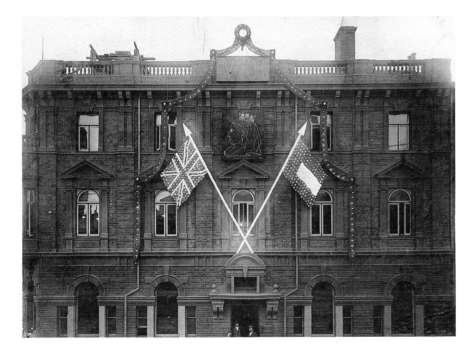

Consolidated Gold Fields offices in Johannesburg, 22 June 1897 – bedecked with illuminations for Queen Victoria's Diamond Jubilee.

miles from Johannesburg, Hammond investigated a working property at the Nigel mine, which had been producing since the rush started in 1886. He got Gold Fields to buy all the claims down dip from that property and in due course this was developed into the great Sub Nigel Mine. The 1908 figures show that Simmer and Jack, with a working profit of £683,385, and Robinson Deep, with £467,875, produced a steady flow of cash during the development of the Sub Nigel Mine, which began to produce the following year.

Yet the history of the Sub Nigel is a cautionary one. Mining geology is more a theoretical than an exact science. It tells you what should be underground; it cannot tell you exactly what *is* there. The gold geology of the Rand is extraordinarily complex. The East Rand has a number of reefs in a mass of valuable 'shoots', running north-west to south-east, broken up by faults. The best of all is the Main Reef, running from surface to 6,000 feet and increasing in value with depth. It took the Sub Nigel men eight years to find the higher grade ore, and during that time the mine only once turned in a profit of over £30,000. It first hit the main shoot in 1917, and profits immediately rose to over £60,000. With some hiccups the profits increased impressively throughout the 1920s. In 1931 Britain went off the Gold Standard, and South Africa dramatically abandoned it too in 1933. The effect was to raise the gold price and to increase the profits of the Rand mines. Sub Nigel made £1,348,728 in the first year of the new gold price system, passed the £2 million mark in 1935 and thereafter made a regular net profit of over £2 million, peaking at £2,825,265 in 1941. Profits fell below £2 million in the late 1940s, then leapt again to £2,299,266 following the simultaneous British-South African devaluation of 1949. Sub Nigel was still yielding handsome profits in the 1960s.

Hence, from the Boer War to the aftermath of the Second World War, the profits of Gold Fields from gold itself depended essentially on three good, long-term mines, Robinson Deep, Simmer and Jack and Sub Nigel, nicknamed Faith, Hope and Charity. In the depths of the Great Depression, 1931–32, Gold Fields paid no dividends on ordinary shares, and without Sub Nigel would have made no profit at all. Big deep-level gold-mines take a long time and a great deal of capital to develop; indeed, they absorb increasingly hefty sums in exploration costs to find in the first place. But once successfully developed they are immensely valuable properties, which last for many decades and can yield uncovenanted extra dividends if political decisions produce a sharp rise in the gold price, as they periodically do.

Nevertheless, any mining finance house seeks to escape the tyranny of gold by diversifying and spreading the risk. Indeed, reducing the risk is one of the house's prime functions. It was John Hays Hammond who first counselled Gold Fields to acquire a spread of investments in the United States. In 1909 the company bought into a series of American light and power companies, a dredging company and the Yuba alluvial gold field. It acquired shares in oil companies in the US and Mexico. Two years later it issued 1,250,000 £1 preference shares to form the Gold Fields American Development Company, which took over the US investments and went in

OPPOSITE ABOVE A military encampment during the Boer War.

OPPOSITE BELOW Recruiting troops for the Boer War at Jamestown.

search of more. Gold Fields also invested heavily in no less than seven West African gold-mining companies, chiefly in the Gold Coast (later Ghana). It took a major interest in the Siberian Lena Goldfields, which eventually became one of the richest alluvial gold finds in Russia, second only to South Africa as a gold producer. It set up a subsidiary company to exploit the potash and borax deposits at Searles Lake in California. From London it accumulated a portfolio of major holdings in chemicals and energy located in the West Indies, Mexico, Romania and a dozen other countries. It began to invest in the gold fields of Western Australia, buying into the Wiluna Gold Corporation and Lake View and Star. In addition to its original holdings in South Africa and Rhodesia, it moved into the South African platinum market, forming with Johnnies the immensely successful Rustenburg Platinum Mines Limited.

By the end of the First World War, Gold Fields appeared to be changing its character. According to the Report of the Directors, 30 June 1918, less than fifty per cent of the company's total investments were in South Africa. In 1918, in fact, the income from its three major mining investments in South Africa was only £72,000. The future was seen to lie in the United States, where a large expansion programme was planned. For technical reasons, a new subsidiary company, New Consolidated Gold Fields, had to be created, making a break with the past. Even in South Africa the balance was switching away from mining. The Managing Director argued in 1919: 'We should turn to a policy of fostering and encouragement of local manufacturers (dependent as little as possible on the importation of raw materials) with a view both to lessening our dependence on overseas industrial conditions and to being able, by creating internal industrial competition, to take full advantage of cheaper local manufacturing costs' (Manager's Report, 17 September, 1919). To this end, Gold Fields acquired

In an attempt to diversify, Gold Fields acquired a spread of investments in the United States in 1909. This adventure occasioned the cartoon 'Piling it on' by Jack Walker, which appeared in *Illustrated Eironee* in 1911.

a brickworks, a furniture company, building and engineering companies, a soap-and-oil works, sugar mills and a rubber factory. By 1920, industrial ventures constituted over ten per cent of its investments. Throughout the 1920s there was a tendency within Gold Fields to take the company away from mining in favour of a worldwide spread of general investments, including manufacturing. It is typical of these years that the company bought into property (Devonshire House), cement and carpet-making. The lacklustre gold price was one reason for this trend. Another was the savage mining strike on the Rand which began in January 1922, led to pitched battles between miners, troops and police, and the loss of 230 lives. Eighteen miners were sentenced to death and four actually hanged. Few now regarded the Rand as a place where quick fortunes were still to be made. So Gold Fields drifted away from its origins and was in danger of becoming a general investment company.

Some of its investments did well. A few were disastrous. Towards the end of the Twenties, as Sub Nigel began to show its tremendous money-making power, gold began to seem more attractive again. Since Rhodes brought the first big mining-pumps to South Africa, the companies with which he was associated, Gold Fields in particular, had always kept in the front track of technology, thanks to men like Tarbutt and Hammond. Even when the company was engaged in global diversification, when its message seemed to be that there was no great future for South African gold-mining, the company's engineers were constantly engaged in daring innovations. In the expansion of Robinson Deep, Gold Fields sank a main shaft to 4,250 feet, easily the deepest in the world in the early 1920s. They then introduced the first sub-vertical shaft, sunk entirely underground from a depth of 3,800 feet, and carrying the workings down to 5,600 feet.

They continued to recruit the best expert talent from all over the world. In 1922 one of the world's most experienced mining engineers, a New Zealander called J.A.Agnew, who had trained under the great Herbert Hoover, joined the board, and the same year the management of Sub Nigel was taken over by a young Canadian, Guy Carleton Jones. Jones enlisted the help of the company's consulting geologist, Dr Leopold Reinecke, in devising a revolutionary approach to the actual working of an established mine. Reinecke had made a two-year study of pay-shoots in ten Rand mines and had concluded that, by careful analysis of the assay results of stoped areas, it was possible to project their extensions in detail into non-mined areas. Jones accepted the idea and applied it systematically to the working of Sub Nigel. Instead of blindly opening up possibly unpayable areas of the mine, the mine-captains and their teams now worked to carefully drawn plans which located the pay-ore to within a few feet. The success of the new system produced a striking fall in production costs and made possible the opening of new mines which otherwise would have been unprofitable: Vogelstruisbult (1933), Spaarwater and Vlakfontein (1934).

Even more important than this, however, was the new confidence in high technology which the success of the Jones-Reinecke partnership inspired. Even at the end of the 1920s, the geology of the Rand was imperfectly understood. The Central Rand had been fairly thoroughly exploited. The

ABOVE Guy Carleton Jones was appointed Consulting Engineer for Gold Fields in 1930. With Reinecke, Krahmann and Merensky he was responsible for the initial development of the West Wits Line.

BELOW Dr Leopold Reinecke, the company's consulting geologist who worked so successfully with Guy Carleton Jones.

East Rand was now being worked. If the basic pattern of the deposits was a kind of saucer, then the gold-bearing reefs must also stretch to the west and south-west and curve round into the Orange Free State. Some were sure of it. But for half a century Renafontein marked the western culmination of the Main Reef group.

In the late 1970s, Dr Dieter Hallbauer, of the Chamber of Mines Geochemical Division, used a scanning electron microscope and other techniques to help elucidate some of the early history of the Witwatersrand gold deposits and the processes of gold concentration. These deposits began to accumulate at the foot of the inner rim of the giant Witwatersrand saucer about 2,700 million years ago, that is halfway back to the formation of the earth. He also showed that gold in some areas was deposited in small plants, among the earliest forms of life on the planet; these formed carpets on the plains and in time became pencil-thick, the height of the plant life. The gold, as it were, collected on these plant-carpets rather as Jason amassed gold by washing auriferous sand over sheepskins, the Golden Fleece. These carpets formed what we know as the Carbon Leader Reef, which is particularly prominent underground west of Johannesburg, sloping down at an angle of twenty-two degrees to a depth of five kilometres below the surface. But none of this was known in 1930.

Men had sought for gold on the far West Rand since 1898, when the Pullinger Brothers sank boreholes on two farms, Venterspost and Gemsbokfontein. They had another go in 1902 when they bought these two farms plus three others, Blaauwbank, Libanon and Uitval. They drilled nine holes and struck payable gold in six of them; indeed, one borehole went right into the middle of what is now the Venterspost Mine, hitting yet another rich reef, the Ventersdorp Contact Reef. In 1910 the brothers started sinking a shaft which they planned to intersect the reef at 2,000 feet. But at less than a hundred feet down they hit water in gigantic quantities – two million gallons a day – and gave up.

Guy Carleton Jones believed in tackling this 'lost' reef and looked for it in 1925. He was over-ruled by expert advice. In 1930, however, he was appointed Consulting Engineer to Gold Fields. This was a fortunate conjunction because in February of that year a German geophysicist called Rudolf Krahmann came to Johannesburg looking for work. His father, Professor Krahmann of the Mining Economics Department of Berlin University, was a pioneer in the use of the magnetometer to detect underground formations. Young Krahmann had also acquired some experience in geophysical prospecting with a magnetometer. In June 1930, while out for the day in the West Rand, he was struck by the glitter of the shales, indicating a high iron content, and found that a tiny piece of shale could turn the needle of his compass. He knew that the structure of the Witwatersrand layers was remarkably regular. The shales of the Lower Witwatersrand were approximately 400 feet below the gold-bearing Main Reef. Hence, as he recorded:

> While playing around with the shale fragment the idea struck me whether it would not be possible to use these ferruginous shale horizons of the

LEFT Rudolf Krahmann using a magnetometer to reveal the riches of the Lower Witwatersrand – 'a tremendous prize'.

BELOW Dr Krahmann in 1932 photographs the future. This was the area from which West Driefontein was to emerge as the king of the West Wits Line – the most profitable gold mine in the world.

Lower Witwatersrand System as a guide to locate the Main Reef of the Upper Witwatersrand System which disappears under younger rocks south of Randfontein.

To put it briefly: if the magnetometer readings could trace the pattern of the iron-rich shales below, the pattern of the gold reef could also be traced, and could then be confirmed by boreholes. Like most brilliant ideas, it was simplicity itself; but no one had ever thought of it before.

Krahmann got some financial backing from Dr Hans Merensky, the discoverer of the great Rustenburg platinum deposits, and cabled his father to send him out a magnetometer tuned to Johannesburg. It arrived at the end of October and by early in December he had confirmed 'with absolute certainty', that the theory worked. He was a passionate German patriot and his initial reaction was to get German firms to exploit the discovery. He turned to Dr Reinecke because he was 'the son of a German missionary', and Reinecke, Carleton Jones and the Manager, Douglas Christopherson, met in the Gold Fields building in Johannesburg. Jones leapt at the idea because he had always believed in the 'lost' reef; Krahmann was immediately put on salary for three months at £100 a month and authorised to spend up to £200 on an immediate survey south-west of Randfontein, in the area of the original boreholes drilled by the Pullinger Brothers.

Krahmann's survey produced a blueprint of the Lower Witwatersrand Series over a huge area of unsurveyed territory, and indicated the faults too. Its importance was confirmed by a leading British geologist, Dr Malcolm Maclaren, who reported it was 'a tremendous prize ... which might well be another Rand'. Credit for the discovery goes in the first instance to Krahmann and to Reinecke for backing him. But the real architect of what became known as the West Wits Line was Carleton Jones, who backed the project enthusiastically from the start – helped by the fact that Agnew, who shared his faith, was temporarily in charge in London, as the chairman was ill – and insisted that the company immediately embark on a major programme of expenditure and acquisitions.

The first big expenditure was £90,000 on eleven boreholes. Eventually twenty-one holes were bored, perhaps the most successful series of boreholes ever undertaken by a mining company, confirming the theory in every particular. Options were purchased on fourteen farms, and Western Areas, a company which held mineral rights over large stretches of the Line, was bought out for £225,000 in cash. To set up such an ambitious programme at the very bottom of the Depression, when the company was paying no ordinary dividend at all, and was only solvent thanks to the profits of one mine, Sub Nigel, required courage. But then, it is one of the salient principles of business history that the true test of an entrepreneur is his ability to take advantage of the downturns in the cycle. The fact that the Rand was profoundly depressed made it a good time to spend money on acquisitions: Gold Fields got the rights over the prospective area remarkably cheaply.

In 1932 it floated the new exploration company, West Witwatersrand Areas Limited, or 'West Wits'. To spread the risk, it offered shares to the other big mining finance houses. Central Mining, Union Corporation and Johnnies turned down the opportunity. Only Anglo-American, General Mining and the Abe Bailey group took up shares. Thus, as another consequence of the Depression, Gold Fields ended up with a larger block of shares than it wished: thirty per cent. It was providential, and was soon seen to be so. The flotation took place in November. Six weeks later on 28 December 1932 South Africa finally followed Britain off the gold standard, freeing the price of gold which surged upwards, setting off a boom in gold shares and extending the life of many existing mines by twenty years.

There was one technical problem of fundamental importance which had to be solved before the field could be developed. The Pullinger Brothers had failed to exploit their original discovery primarily because the rush of water at around 100 feet made shaft sinking impossible. The West Wits reefs lie under a mass of porous dolomite rocks. Seepage of rainwater has hollowed out vast caverns within these rocks which fill with water and thus constitute enormous underground lakes poised above the gold-bearing conglomerates. To mine the gold it is necessary to drive shafts through this giant rocky sponge, which continues to constitute a water hazard to those below it throughout the life of the mine (as we shall see). Happily, the discovery of the West Wits Line virtually coincided with a new development in excavation technology. In 1920, through the initiative of J. A. Agnew, Gold Fields formed links with a Belgian company called Francois Cementation. It specialised in injecting a cement mixture, known as grout, into rock fissures to seal off water. Cementation was brought into the West Wits project at an early stage and devised a method of cementing up the dolomite fissures surrounding the shaft area at each stage before lengths of the shaft were sunk. Thus the shaft passed safely, though not without considerable trauma, through the dolomite in its concrete cage.

Truter, who was a young geologist working for Gold Fields at the time, summed it all up for me: 'The West Wits Line had been dormant for thirty years because of the flooding. Then geophysicist Krahmann, aware of the fact that certain magnetic beds occur at a known constant distance from the reef, demonstrated that with his magnetometer he could trace the magnetic beds, and hence the reef, even under the dolomites. This generated a renewed interest in the question of reef extensions under younger beds. But what about the water? Gold Fields had connections with Cementation. So we had a coming together of two new technologies. Gold Fields introduced geophysics into South African mining, and it introduced the new technology of cementation.

'Krahmann followed the reef far beyond its previous known limit. The reef had changed direction, but he could cope with this using his magnetometer. There were the right people around at the time – Dr Reineke and Mr Carleton Jones. They recommended to go ahead in spite of the Depression. This was then the biggest single exploration project undertaken by Gold Fields. They acquired rights over fifty miles, the major portion of the gap between two outcrop areas. You get mineral rights cheaper in a Depression. Gold did not go up to thirty-five dollars an ounce until 1936, by which time we had already delimited four mines. The whole thing was very bold. Take the drilling programme – at the very start a dozen holes were sited simultaneously, and several old boreholes were cleared out and deflected to obtain fresh samplings of the reef. Very bold. It all turned out exactly right. The gold field was there. The cementation process worked, though not too well to begin with. The first two shafts of the first new mine were near the old shaft, and the same water trouble came but this time we got the cementation process to work and the shafts were saved. That was Venterpost, the first of the new mines. It began producing in 1939 and it's still going.

Anglo-American got into West Wits because Gold Fields had decided to limit mining to a maximum depth of 8,500 feet. Anglo went for the deeper part of the reef lying between 8,500 and 13,500 feet. This is the reef known as the Carbon Leader. It is an extension of the principal reef of the Central Rand, the Main Reef. They were exceedingly lucky when it was found that at a much shallower depth there was another reef nobody had ever heard of in this area. This important reef, among the richest, was at the bottom of the lavas. Gold Fields had identified this reef at Venterspost, and it is known as the VCR, the Ventersdorp Contact Reef. All but two of the West Wits mines have it. Johnnies also got in and established themselves in an area where the reef might have been too deep but has been lifted to a shallower position through faulting. 'Fifty years later we are still studying the West Wits Line, and we have learned a lot about its reefs. Every time we thought we were nearing the end of things we have found new prospects. So we have developed immense confidence in enlarging the areas on which we are working and so extending the life of the mines.'

Truter's enthusiasm for this great gold-field is understandable. It is the pearl of the Rand. But developing it took a great deal of money and time. Between the discovery of the field in December 1930 and the launching of the first mining company, Venterspost, on 5 June 1934, there was a gap of three-and-a-half years. The Number One Shaft at Venterspost met water-rushes of up to 600,000 gallons an hour, at 112 feet down, 200 feet and 450 feet: without Cementation it could not have been sunk at all. The mine did not begin milling until 1939, so it took the best part of a decade before the first gold was produced. Indeed, West Witwatersrand Areas Ltd. did not pay its first dividend until fourteen years after its flotation. The second West Wits mine, Libanon, was launched as a company on 4 August 1936. It was expected to start milling in 1942 but the war forced its suspension. The third mine, Blyvooruitzicht, was in an area where Gold Fields had not been able to acquire all rights: a strip was held by a company ultimately controlled by Central Mining; so West Wits Areas, holding some eighty per cent, joined with Central Mining to establish a mine. This mine began milling in February 1942 and was the first mine to show payability over the whole of its area, from one boundary to another.

However, the king of the West Wits Line, West Driefontein, had its development held up by the Second World War. It is situated underneath the edge of a gigantic underground ocean, and if it is the richest it is also the most water-bedevilled of mines. Sinking its Number One and Number Two shafts tested all the resources of Cementation, and it was not until 1952 that it first began to mill gold. By that time, Gold Fields had invested more than £20 million in developing the Line. But in the 1950s and still more in the 1960s it was able to reap a spectacular financial harvest. The effect on profits of the 1949 joint British-South African devaluation was dramatic, above all on the low-cost West Wits mines. The working profit from Venterspost jumped from £556,394 to £1,290,044 in 1950. West Driefontein, in its first full year of production, 1952, turned in a working profit of well over £2 million, which by 1955 had doubled, doubling again by 1958. This was the first mine in the world to make a working profit of £1 million a month and by the end of the 1960s its annual profits were over £20 million. It and its partner East Driefontein (worked separately, but divisions of Driefontein Consolidated) have broken all records in mining profitability and in 1985 it was still the lowest cost mine in the industry with a further projected life of fifty years. Depending on the current price of gold, its capital value has been estimated at between £2 billion and £4 billion.

The return on capital thus bears comparison with the best days of the Central Rand in the 1890s. However, developing the West Wits Line not only taxed Gold Field's financial resources to the limit but, a point often overlooked, it absorbed a high proportion of the time and energies of its top administrators, geologists and mining engineers. This helps to explain why Gold Fields had less success in acquiring and developing properties at the western extremities of the Line, and in following the gold-bearing rim of the saucer across the River Vaal and into the Orange Free State. It also explains

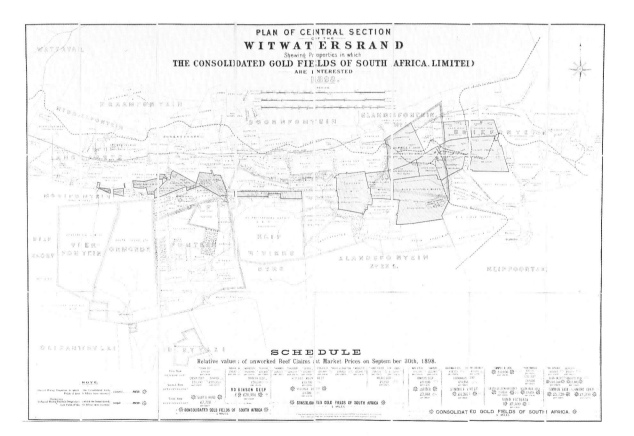

why it did not become the Number One mining finance house in the industry.

The mining finance houses are *confrères* in the sense that they often hold shares in each other, and in each other's mines, and from time to time engage in joint ventures to spread the risk. But they are also rivals and mutual predators, capable of engaging at short notice in atrocious acts of cannibalism. Although Gold Fields was the first of the houses, the leading house by the end of the 1890s and for long after was Central Mining, incorporating Rand Mines; it held this position primarily because Beit was more interested in gold than Rhodes and therefore better at finding it. Gold Fields held the second position and has continued to hold it through all the mutations in the structure of the industry. At the top, however, the position began to change from the 1930s with the emergence of Anglo-American Corporation of South Africa, a company formed by Ernest Oppenheimer in 1917 with the assistance of Colonel W.B.Thompson of the American company, Newmont Mining. The purpose of the new company was to enable the Oppenheimer interests to draw on the New York capital market. But its financial strength was enormously increased by the success of Oppenheimer, in the course of intense and brilliant manipulations lasting from 1920 to 1934, in securing through De Beers (which he acquired) the unification and control of between eighty and eighty-five per cent of the entire world diamond market. Simultaneously, another of his groups was establishing a dominant position in the Rhodesian copper belt. In the 1930s and the 1940s, Anglo-American moved decisively into the new gold fields of the Far West Rand, Klerksdorp and the Orange Free State south of the Vaal.

Oppenheimer and his associates were not particularly good at finding minerals but they were unusually skilful, energetic and persistent in acquiring companies which already had them. They also had much greater financial resources than Gold Fields. The Blyvooruitzicht results, from 1942 on, showed that the Carbon Leader became richer with depth and as it extended below 8,000 feet, mining at levels never before worked became justified. In addition to its other ventures, Gold Fields did not have the money to embark on these ultra-deep level risks. It eventually took a substantial share in Western Ultra Deep Levels but it was Anglo-American which formed and controlled the exploration company and then created the mine, sinking main and sub-shafts to 10,000 feet and investing nearly £30 million before the mine began to mill.

At the same time Anglo, with Anglo-Vaal and the Strathmore Company, were developing the Klerksdorp Mines. In the Orange Free State, it was clear by 1939 that the Basal Reef occurrence there constituted an enormous gold field. It was then controlled by African and European Investments. The efforts of Gold Fields to secure an interest in this company failed. It also turned down Oppenheimer; but by ferocious share-buying he managed, over the years 1939–45, to penetrate and finally control the company, thus securing the future President Brand, Welkom and President Steyn mines. Gold Fields joined in the Orange Free State prospecting hunt late – it had its hands full elsewhere – but in the end it tried very hard, taking and prospecting options over nearly 2,500 square miles. It had no luck. It ended up at the end of the scramble with one dud mine and an interest in the successful Harmony Mine.

Hence in the decade after the war, even at a time when the West Wits Line was coming to full fruition, Gold Fields still found itself Number Two to the new giant, Anglo-American. The position was an uneasy one because of Anglo-American's propensity, backed by unrivalled financial resources, to worm its way into all its competitors. One historian of the industry, C.S.Mennell, noted in 1961: 'Since the end of the war [Anglo-American] has become not only the predominant group but in many ways the leader of the industry. Its great financial strength has enabled it either to control most new ventures as they emerge or to acquire very considerable interest in such ventures . . . Anglo-American is represented on the boards of twenty-one of the twenty-three gold-mining companies floated since the war ... In addition this group has its representatives on the boards of no less than three of the other seven mining groups ... Not only has Anglo-American become the undisputed leader of the industry, it has also become, to a considerable extent, the banker of the industry.'

By 1968, after the opening of Western Deep Levels, Anglo's twelve mines controlled forty per cent of the gold produced by the industry, representing three per cent of South Africa's Gross Domestic Product. In 1975 the score was: of thirty-nine mines altogether, Anglo-American controlled twelve; Gold Fields and Union Corporation seven each; Barlow Rand five; General Mining four; and Anglo-Transvaal and JCI (Johnnies) two each. Gold Fields tried to strengthen its position, both relatively and absolutely, by gaining control of another house. In 1955 it made a tentative attempt to merge with Central Mining. Anglo forestalled it, and acquired a controlling interest. In 1974 Gold Fields tried to gain control of Union Corporation, but this attempt also failed. The Oppenheimers had long been anxious to satisfy the natural desire of Afrikaner financial interests to have their own major mining finance house. In the 1960s the Afrikaners began to nibble at Johnnies. This the Oppenheimers would not permit, as Johnnies had thirty-five per cent of its assets in diamonds, with a large holding in De Beers and the Diamond Producers' Association, one of the ultimate sources of Oppenheimer power. So their primary holding company, Rand Selection Corporation, promptly took a fifty per cent stake in Johnnies. By way of compensation, they assisted the Afrikaners to take over General Mining, and after the Gold Fields bid for Union Corporation, the Oppenheimers supported a move by General Mining which led to the creation of General Mining-Union Corporation, or Gencor as it is known, in which Anglo-American had a substantial minority interest. In the period October 1979 to February 1980 Anglo-American bought shares in Gold Fields itself, acquiring a holding of twenty-nine per cent.

By the early 1980s, in fact, Anglo-American controlled Johnnies, and it had large minority holdings in Gencor, Gold Fields of South Africa and Barlow-Rand; only the smallest group, Anglo-Transvaal, was outside its orbit. In theory, of course, Anglo-American itself is vulnerable to predation. But it is indeed a conundrum, since the four inner holding companies of the group have a complex series of cross-holdings in each other. Each of these is itself a minority holding but added up the four control every company by large majorities. And the size of the whole is a daunting obstacle. As the Johannesburg *Financial Mail* put it, 'Any international giant attempting (a takeover) on a share-swap basis would probably end up in the uncomfortable position of having Oppenheimer as its biggest single shareholder.'

This disquisition on the relationship between Gold Fields and Anglo American has been necessary partly because Anglo is the largest single shareholder in CGF but also because the inability of Gold Fields to become the undisputed leader of the Rand helps to explain its quest for wider horizons in the 1950s and after. It was not, of course, the only or the chief motive for diversification. As we have seen, there was a period of rapid diversification between 1909 and the mid-1920s. Although managements can always find compelling reasons for these major changes of policy, the fact is that fashion has a great deal to do with it, and the need of managements not only to be active but to be seen to be active. Businesses tend to evolve in cycles. Success is achieved and large profits made in developing a single product, in CGF's case South African gold. Management then becomes worried at the risks inherent in this over-concentration, and diversifies. The next phase of the cycle begins when the results of diversification prove financially disappointing, and management decides that the true role of the company is to do what it really knows about and focus its efforts on the original product. So then there is a process of re-concentration and a new cycle begins.

In the 1950s there were particularly cogent reasons for spreading CGF's assets more widely. For the past twenty years the company had concentrated on its main product and the results were spectacular. The 1959–60 figures were the best in the company's history. Gold Fields was now producing one-tenth of the free world's gold. But the overwhelming confluence of assets and profits on one product in one country – over eighty per cent in Rand gold – made it mercilessly dependent on the gold price, a factor quite beyond its control. In 1951, Sir George Harvie-Watt, a wartime Parliamentary Private Secretary to Winston Churchill, with a background in law and commerce, became managing director of Gold Fields, Deputy Chairman and Chief Executive in 1954, and Chairman in 1960. Throughout his period of power, with the enthusiastic support of his prime deputies, Gillie Potier, Donald McCall, Robin Hope and Gerry Mortimer, Gold Fields underwent vigorous diversifications, both in products and geographically.

Within South Africa, acquisitions enlarged the spread of its products. In 1957 Gold Fields was the senior partner, the Chartered Company and Anglo American being associates, in a takeover of the South West Africa Company, with Gold Fields assuming its management. In 1959 the company took over the Anglo-French Exploration Company, which gave it control of Rooiberg Minerals and Apex Mines, and interests in oil, coal, copper and tin. In the same year it acquired New Union Goldfields, and with it Union Tin, Star Diamonds and Glenover Phosphate. To administer this new spread of investments, CGF acquired the assets of its operating company, New Consolidated Gold Fields, and subsequently, in 1972, West Wits was absorbed into Gold Fields of South Africa (see page 54), which then became a quoted stock on the Johannesburg and London stock exchanges. Meanwhile Consolidated Gold Fields (of London) had dropped the 'of South Africa' from its title in January 1964.

Spreading the risk within South Africa was only part of the programme. In 1948 the Nationalists had come to power in Pretoria and immediately set about implementing their philosophy and programme of apartheid. During the 1950s the anti-colonial revolution gathered pace; 1960 was the year of the 'Wind of Change' when the former black colonies, in rapid succession, began to attain independence. In 1961 South Africa left the Commonwealth and became increasingly isolated on the world scene. There was a feeling then, which persisted until the gold-price revolution from 1972 onwards, that South Africa had only a limited future as a white-run state and that it was not a premium area for long-term investment.

Hence Harvie-Watt's decision to again embark Gold Fields on global diversification was more than fashion; it seemed at the time unavoidable prudence. The overseas thrust was in three main directions. First, there was rapid expansion of investment in Canada involving the creation of an exploration company and a Toronto-based mining finance house, Newconex Holdings Limited. Next, from 1960 onwards, Gold Fields began to acquire new properties in Australia, where it already owned working goldmines in Kalgoorlie and elsewhere. Through a new company, Consolidated Gold Fields Australia, it secured a big interest in the mineral sands industry, iron-ore deposits in Mount Goldsworthy in the north of Western Australia, and control of the Mount Lyell Mining and Railway Company, the oldest and one of the most famous copper mines in Australia.

Above all, it expanded in the United States. Its efforts centred on the US zinc industry. It bought a ten per cent interest, then in 1963 outright control, of the American Zinc, Lead and Smelting Company, which produced about thirteen per cent of US slab zinc and was its second largest zinc concentrate producer. This purchase, which cost the then enormous sum of $18 million, was Gold Fields' largest single investment to date. It was the climax of a programme of overseas ventures which brought the company some twenty new firms at a cost of about £20 million, and raised the market value of its assets to £80 million. Within this global figure, by the mid-1960s, South Africa accounted for fifty per cent, North America for twenty-five per cent, and Britain and Australia for the remainder.

The recent development and the current prospects of Gold Fields' interests in South Africa, North America, Australia and Britain will be analysed in later sections of this book. Here we are concerned with the main strategic thread. The trouble with diversification, as many large and well-run companies have discovered to their cost, is that unless it proceeds according to a definite and clearly thought-out plan, based upon a coherent philosophy which answers the question 'What are we in this particular business for?', it is liable to develop a logic of its own and carry the parent company, a helpless spectator, into strange and dangerous places. Gold Fields' American programme was a case in point. When buying a company, the purchaser should not ask himself 'How can I handle the upswing?' –

anyone can do that – but 'How can I handle the next downturn?' which will follow as surely as night follows day. The question is particularly vital in the world of commodities, where the swings are spectacularly fierce and dramatic, and in manufacturers based on commodity sales. In the decades 1960–1980 Gold Fields did not ask this question often enough or loudly enough.

With the purchase of American Zinc, Gold Fields for the first time became a member of the big-league US mining industry. It came in on the upswing. Before the purchase, American Zinc's profit had been £672,000. By 1965 it had risen to £1,698,000. In 1966–67 came the downward plunge in the price of zinc, aggravated by a disastrous strike. One of those who was around at the time, Vincent Filippone, now a Vice-President of Gold Fields American Corporation, described to me the sequel: 'In 1968–70 the losses were so great, because of the strike and low zinc price, that Gold Fields, having spent $30 million on an expansion programme and a new zinc refinery, was forced to sell its best asset, its zinc reserves, the best in the United States. It also sold off its refinery in St Louis. Then zinc made a comeback but for Gold Fields it was too late. It was left with assets worth only $1.5 million and a $44 million tax-loss.'

That tax-loss set off a further logical chain of events of its own. To use up the tax-loss, it seemed sensible to acquire some profitable companies. The logic pointed in the direction of manufacturing to acquire as Filippone put it to me, 'sure, reliable manufacturing companies, making steady profits and, since it wasn't Gold Fields' business, with easy-to-see assets'. The man increasingly in charge of the new American operation was David Lloyd-Jacob. Gold Fields, of course, had been in manufacturing before, in the period 1918–22. As a result of the Harvie-Watt expansion programme in South Africa, and the acquisition of H.E. Proprietary, it took over two British metal-manufacturing companies, Alumasc Limited and Metalion Limited.

The new programme was on an altogether different scale. It continued throughout the decade of the 1970s and involved acquisition of a number of firms in the steel industry and energy-equipment fields. Most of these purchases were shrewdly calculated. They made substantial, in some cases handsome, profits, thus fully justifying the tax-loss strategy. But as the decade proceeded the original object of the acquisition programme was forgotten. Businesses were retained after the tax-loss was exhausted; more were added. An alien logic took over completely. The climax came in 1980–2 when an oil-rig manufacturer, Skytop Brewster, was bought, made record profits in 1981, then nose-dived as the oil exploration boom collapsed. In 1982–3 Gold Fields was forced to meet losses of £89 million on this one company. Lloyd-Jacob resigned, and the thrust into manufacturing was abruptly halted.

Meanwhile, however, the elements of a more sustainable long-term strategy, springing from a genuine Gold Fields' business philosophy, had begun to emerge. In 1981 Gold Fields bought a seven per cent interest in Newmont Mining, one of the largest mining groups in the US; the stake was later to be raised to twenty-six per cent. This was an ambitious investment, involving heavy borrowing and deployment of a very large percentage of Gold Fields' total resources. One of Newmont's main products was copper. By the mid-1980s the prolonged downturn in copper prices made the investment look expensive. On the other hand, Newmont has an impressive variety of reserves, and in the long term the involvement with this company accords well with Gold Fields' policy of acquiring a global spread of reserves over a wide range of minerals. At the same time, building on an investment going back to 1968, Gold Fields had been expanding the Amey Roadstone Corporation, first in Britain, then in the United States, with a multitude of properties in the construction materials' business. Steady if undramatic expansion in this field had made ARC, by the mid-1980s, one of the most successful and profitable elements in the entire group.

These developments enabled John Agnew's grandson, Rudolph Agnew, who became Deputy Chairman and Group Chief Executive of CGF in 1978, and Chairman in 1983, to devise a working philosophy for the company which took account of its existing assets and provided it with a plausible set of objectives. It was based on two propositions. The first was that Gold Fields is essentially a company engaged in finding and exploiting natural resources, from gold to sand and gravel, with a manufacturing capacity only where this is relevant to the primary object. The second is that it is a commonwealth, not an empire, and the function of CGF in London is to provide the financial and philosophical framework within which each component can individually pursue its manifest destiny. We will now examine how this system functions across the globe.

2 South Africa: Achievements, Problems and Opportunities

THE ORIGIN OF CONSOLIDATED GOLD FIELDS, and still at its financial core, are the gold mines of the South African Rand. It is true that in 1984, for the first time, the biggest single contribution to Gold Fields' profits came from outside South Africa, from the Amey Roadstone Corporation in Britain. South Africa provided forty-five per cent of the profits. But in some ways these statistics are misleading, and the contribution from South African gold in real terms is well over fifty per cent. Antony Hichens, a Managing Director of Consolidated Gold Fields, explained this anomaly to me:

> It is due to an accounting quirk. When a mine has been discovered and is about to be developed, it is normal for a mining finance house in South Africa to sell part of the share capital to the public. Deep-level mining is hugely expensive and this spreads the capital risk. When the mine starts to earn a profit it will distribute it in full to its shareholders after paying tax and meeting necessary capital expenditure for further development. Conventionally, the mining finance house will take only this dividend into its accounts, not its share of the mine's pre-tax profits. Thus our share of Gold Fields of South Africa's profit, and the profit from our direct interest in some of the Group's South African mines, show in our accounts as after-tax dividends, whereas with a wholly owned subsidiary we show its pre-tax profit.

The confederate nature of the mining finance house structure has now been extended upwards to Consolidated Gold Fields itself, so that London no longer has an absolute controlling interest in the South African house. For the past fourteen or fifteen years it has ridden its principal asset with a light rein. 'The present worth of Gold Fields of South Africa', Hichens told me in early 1985, 'is about £1.3 billion, with assets of £1.5 billion. So it is a very big company in its own right. Moreover, GFSA, being a big company, does not feel the need for detailed guidance from London on its operations. They have 700 employees at their central office in Johannesburg, a far bigger staff than Gold Fields itself. They are knowledgeable about politics. They have great technical strength. Having once been a mere agency of London, they are today an independent mining finance house. Once it was London which placed all the major machinery orders. And at Gold Fields' headquarters in Moorgate there was a major consulting engineers' department. But this was becoming more and more of an anachronism. So in the mid 1970s the board cut all this out. At the same time it changed its relationship with GFSA. Once it was a wholly owned subsidiary of Gold Fields. Then in 1971–72, GFSA and the West Wits company were amalgamated, and CGF ended up with forty-nine per cent. So GFSA is now as independent a company as you can have, granted a single major shareholder. I think it is better that it should be independent. But it must share decisions with Gold Fields on three points: major new business projects, dividend policy, and the appointment of its chairman. What all this means in practice is that Gold Fields is a company with a major asset which is substantially independent. We might not get a good South African chairman if we exercised our power too frequently. This has been the position over the past decade and more.'

The consolidation of the current status of Gold Fields of South Africa, answerable to all shareholders, is naturally viewed with satisfaction at GFSA headquarters in Fox Street, Johannesburg. Robin Plumbridge, its Chairman and Chief Executive Officer, reflects:

At one time we were a branch office. Then we moved to a new phase in 1971–72 and were deliberately created as an associate. What has happened in the last four to five years is to define our relationship with London. Rudolph Agnew, Chairman of Gold Fields, has done a superb job. During this time we have been through a period of phenomenal growth in market value. There has been worldwide recognition of us as an independent unit and financially we have become a very powerful company. We have minimal indebtedness, so we have our own capacity to take advantage of opportunities as and when they occur. Has London come round to GFSA philosophy? What London has come round to is the view that the child is now an adult in its own right. After all, there have been times when London itself was not sure of its own identity, and in particular its identification with the South African company. That recognition of adulthood is not an easy process. But we now talk to each other as mature people. We debate policy issues. We talk to each other as equals, and that is what the confederation is all about.

But if Fox Street now takes for granted a wide measure of independence from London, it concedes a corresponding (if more limited) degree of freedom to its associate mining companies. As Plumbridge put it to me:

We have a clear-cut basic philosophy and a clear-cut structure which supports that philosophy. We have always, out here, worked on an associate basis, with holdings below fifty per cent in each mine. We have a *de facto* control rather than absolute local control. We have always had this. We provide technical services, but we don't seek to dominate. We are easily embarrassed by a totally owned situation. All this means that the chairmen of the individual boards of our mines are publicly accountable. They stand up once a year to be shot at by shareholders. We thus have to be seen to be acting in the best interests of that particular company. We can't use their resources for the benefit of a particular shareholder. This is a very good discipline. It means that the interests of the individual company come before those of the group. We apply it to a rather greater extent than other mining finance houses, though all are composed of publicly quoted companies. It is a characteristic of the South African mining finance system.

Where we differ is that, no longer having a manufacturing sector, we don't run into conflicts of interest between mining and manufacturing sectors. For instance, we can choose all our own equipment – we don't have to buy from our linked firms. When we engaged in industrial activities this was a continual bone of contention. It caused a huge waste of executive time. Now it all must be done on a competitive basis. We don't grant preferences to anybody. The concept of association, with the public participating, was present in South Africa right from the beginning. And it suits us very well.

Gold Fields of South Africa now has a spread of investment in a wide variety of products. As Plumbridge says. 'Some ten years ago we rededicated ourselves as a natural resources company.' But as with Consolidated Gold Fields itself, gold is its heart. GFSA prosperity is based upon seven gold-mines, including the richest and lowest-cost producer in the world. I was given an analysis of the structure and operations by Colin Fenton, GFSA's Executive Director in charge of gold activities:

Until 1932 Gold Fields was operating only on the Central and East Rand. Now all our underground mines are on the West Rand, though with the high price of gold we are reworking our old surface dumps in the East. We began our first West Wits mine in 1934. That was Venterspost. It's now fifty years old and still running. Not a rich mine. Currently it is operating on a break-even point. On the other hand, if the gold price doubles, its profitability multiplies four to five times.

Then there is Libanon. It started in 1939 but was on a care-and-maintenance basis during the war. It went into full-scale production only in 1951. In 1967, with gold still at only thirty-five dollars an ounce, it had a possible life of only eight years. When I was running it, I was told to mine out the shaft-pillars, the prelude to closing down a mine. Then the Nixon policy change resulted in a higher gold price and that extended its life. It has still got thirty years today.

Next came West Driefontein in 1952, Doornfontein in 1954, Kloof in 1968, East Driefontein in 1972 and Deelkraal in 1980. In 1982 we consolidated East and West Driefontein and the new North Driefontein area in one company, but East and West continue to be worked as two separate mines.

Gold-mining is a peculiar business in certain respects. Those who own and work the gold-mines have no control over the price of their product. In South Africa there is only one customer, the government, to whom all gold must by law be sold. But the South African government has little control over the gold price either, which is subject to world fluctuations. When people lose faith in paper currency, they move into gold; when they regain faith in paper currency they move out of gold. So gold tends to be the mirror image of the prime paper currency, the dollar. A weak dollar pushes the gold price up; a strong dollar pushes it down. But the profitability of South African gold-mines depends not merely on the world price of gold, expressed in US dollars, but on the relationship between the South African rand, in which the government pays for the gold it buys, and the US dollar.

OPPOSITE Surveyors near the site of West Driefontein in 1933.

56

Gold reached a peak at the beginning of the 1980s of $800 an ounce and even more; it has since drifted down to the $300–$400 band. But, as Fenton explained to me, the 1984–85 fall in the rand against the US dollar meant that GFSA profits from gold were higher:

The fact is that gold prices are higher than ever in terms of the rand – up to thirty-five per cent on 1983. As we are paid in rand and incur costs in rand, it is the rand price of gold, rather than the dollar price, which matters to us in the short term. It is now R19,500 a kilogram – a year ago it was R14,444 – so the gold division is smiling everywhere. Of course one has to make allowance for twelve per cent-plus domestic inflation, and the fall of the Rand against other currencies. But virtually all materials GFSA buys are locally produced, so we are in good shape.

There are some curious Alice-in-Wonderland aspects to gold-mining economics. The mine-owner cannot control the price of his product. At the same time, his underground assets are limited and his end product varies in cost of production.

Gold ore in the ground is defined as 'pay' or 'unpay', depending on whether it is profitable to mine or not. But the pay definition varies with the price of gold. So, as Fenton puts it:

The higher the gold price, the less gold we have to go for. Let me explain. In South Africa, you can split land and mineral rights. First you get an option from the mineral right owner then from the government a permit to prospect on the farm. A borehole can take years. You may have a three-to-five year option. Often you walk away after five years. Or you may exercise your option to buy the mineral rights, if the prospects look good, or even the surface rights too, that is the freehold. If your feasibility study is favourable you go to the government and apply for a mining lease, because in the end all precious metals belong to the State. This means you must propose to and negotiate with the government a payments' formula.

He showed me the mining lease application, dating from 1974, of Deelkraal, the new mine which began milling in 1980.

This lease shows the payments' formula. It lists the owners of the land – there may be hundreds of them, each with a tiny bit. It gives details of the borehole results and so forth. It describes how we proposed to sink the shaft and how we would mine the ore. It gives the time-scales for the period 1982–86 and thereafter.

Eight times out of ten, government accepts the formula we propose, which is in effect a percentage of the profits. But the government takes two bites. One is on the lease, usually referred to as the state's share, which the mine treats for tax purposes as part of its costs. The second is the tax on profits. The total the state may take from a rich mine may be over seventy-five per cent. But from a poor mine it may get nothing.

Once a mining company has a government lease, the farmer who still hangs onto the freehold is in a weak position. Under the gold law, in

OPPOSITE ABOVE Venterspost in June 1934. The first West Wits mine, now over fifty years old and still going strong.

OPPOSITE BELOW The offices at Venterspost in 1939.

Venterspost today.

terms of which the lease is granted I can put my head-gear right in the middle of his farm. Only his homestead, water-borehole and developed lands and shed, known as 'Owner's Reservations', are protected. A Surface Rights Permit is issued by the Mining Commission, and this means I can cover his beautiful farm with slime. He is, however, paid a rental, and it is generally beneficial to the company to purchase his property.

Right: the next thing is to sink the shaft. The mining plan is shown in the lease, and mining is to some extent subject to government supervision. Basically, you have two shafts, one out, one in.

Each mine has a mine plan, which is constantly updated, and a new one issued every year for the information of shareholders. The reef is like jam in a sandwich. The jam is the ore reserve, marked yellow on the plan.

This indicates what is to be taken out in the next year – one year's milling. Worked out areas are marked in solid colours for reef such as blue, red or green. There may be fifty years' reserves of ore in a mine, but the Ore Reserve is only for one year – the company values only what it can mill next year. In terms of your lease you are obliged to mine to the average value of your Ore Reserve. This means bad as well as good grades. However, when the gold price fluctuates upwards, rising say from $300 to $600 an ounce, this means you can mine what has been unpay ore, which now becomes payable. So as the gold price goes up, we mine lower-grade ore and produce less gold but for no less profit. That, for instance, is why we and others are mining the old Johannesburg dumps, to get gold out of what was once discarded as unpayable. And with the gold price high we can bring mines back into production.

Take Kloof, a normal mine. With the gold price high, you produce 800 kilograms instead of 1,000. You get the same money for less gold, and you are mining your assets to the full. Production costs are expressed in cost per ton mined and/or milled. It is cost per rock ton, not cost per gold ounce. The 1985 mining plan of Kloof shows the state of the game. The blue shows sixteen years past mining. The yellow shows next year. What is left is white – forty years mining.

You have to mine to the average of your ore reserves. But with the gold price changing daily, your pay limit and your average ore reserves change daily too. Mining is not wholly unlike farming. It is governed by prices, just as the price of plums, say, determines the farmer's profit. The only thing you can control in gold-mining is your costs. We work on a six-month moving average, to iron out fluctuations. You work out the pay limit for the six-month average price. Everything above is pay. Everything below is unpay. The reef value, say, is ten grammes. The value from the stope, or face, is nine, because of waste. Then you have to get the rock out. If the gold price is low, you put it on the dump. If it is high, you send it to the mill. Stope dirt at ten grammes per ton goes to the mill. But mistakes are made. Sometimes stuff goes to the mill at nought grammes per ton.

The value of the ore is estimated from boreholes. Some twelve

ABOVE Libanon – shaft sinkers in transit.

LEFT Libanon – drilling at the mine face.

boreholes, at the lease stage, are often enough to risk a capital investment of a billion rand on which there probably will be no return for at least seven years. Thereafter there must be a process of continual sampling. If your sampling is perfect, and the mining is perfect, you get the ideal figure of millable ore out. But there is a margin of error in both cases. The aim is to reduce the margin. On this basis you can move your targets daily, and Gold Fields was the first to do it. According to the pay limit you say (for instance), 'Mine at six-and-a-half grammes per ton.' You then move men to lower-grade faces, as the gold price (in rands) increases. You are aiming at a moving target, so computers are immensely valuable.

However, all generalisations about gold-mines are liable to be misleading. No two mines are alike. Each has a history and life of its own. However accurate the borehole sample may be, no one knows what is really underground until the mining starts. The mine's development, too, is affected by variations in the gold price, by government policy and by the availability of company resources for investment. Robinson Deep, Simmer and Jack and Sub Nigel all lasted for decades longer than the men who first designed them had expected. Robinson Deep finally stopped mining in 1964 and was sold outright. Simmer and Jack was sold in 1965; it is still producing, though at minimal profit. Sub Nigel was closed in 1970 and turned into a property company.

The development of the West Wits line was also subject to luck, hazard and government decisions. A former GFSA deputy chairman, Peter van Rensburg, an expert on valuation, described some of the turning points to me:

Libanon had been regarded as a low-grade mine. Then they began to strike the higher grade VCR. That in turn led to the proving of the area that became the Kloof mine. At a time of financial stringency the initial recommendation was to develop the area as an extension of Libanon to make use of the latter's tax cover. We had already forecast a gold price increase, but, of course, we didn't know when or how much. Unfortunately we could not agree among ourselves on a valuation that would be fair to both the shareholders of Libanon and the holders of the mineral rights of the Kloof area. Too optimistic an estimate would favour one party and too pessimistic an estimate the other party. Reluctantly it was decided to develop a separate mine but buying services in the form of predevelopment from Libanon. We had to skimp in developing Kloof. More money could have made a better mine. But we had only some R40 million whereas about R70 million would have been more appropriate. With more money we could have had another hoisting shaft and got started on the subvertical shafts. So we were limited by the amount of stope face we could open up on the single vertical hoisting shaft. Kloof has better ore even than Driefontein, the best mine in the world, but it can't get as much ore out.

East Driefontein was another interesting case. It shows it sometimes pays to re-drill an area when previous drillings produced no values. After being overlooked twice before, it was reassessed and has proved to be a winner. But because of the previous doubts we wanted to develop the East Driefontein mine as an extension of West Driefontein, similar to the proposal for the Kloof area. A new scheme had been developed to protect both sets of shareholders. But the Government Mining Engineer, as Chairman of the Mining Board, refused to give his permission. The government would have had to wait for its tax income. So it had to be two mines. Sometimes boreholes mislead in the other direction. At our Free State mine we misinterpreted the borehole values. The reef turned out much more patchy than we expected. The mine was much less profitable than we'd hoped, it had to borrow money, and the interest on borrowings soon absorbed what profit there was. So in the early 1960s we sold it to the neighbouring Anglo mine which took it under its tax umbrella. In the same way two other mines of the Anglo Vaal group were amalgamated into its southern neighbour, the Harmony mine of Rand Mines.

From March 1972, as a result of President Nixon's change of policy, the gold price began to move up. Our geologists had no difficulty in persuading the board to examine an area south of Doornfontein, which has since been opened up as the Deelkraal mine. But it has not so far been a huge success like East Driefontein and in fact is not as good as we'd hoped. To prevent the shaft-pillar immobilising good ore, we put the shafts in a low grade area. We are only just approaching deeper areas where, on the basis of boreholes from surface, higher values are expected. The Anglo mine next door is similar – now running into some better areas. But it may be a bit patchier than the boreholes show. We shall see.

Each mine has a character of its own. Adriaan Louw, Chairman of GFSA (1965–80) told me:

Mines can be rich, clean, painted, modern tools and toys. But they can also be happy when they're poor because people care for them. It is usually done by one person, a good manager, who creates the spirit of the mine. Colonel S.R.Fleishcher put his spirit into Sub Nigel, which he once managed. That was a happy mine. The spirit of West Driefontein was implanted by Stan Gibbs, a protégé of Fleishcher. That was and is a happy mine too. The two years I spent as manager of West Drie was the best time of my life – to stay there I almost turned down promotion to a head office job.

West Driefontein is the wonder of the Rand; indeed, of the gold-mining world. It has had its share of problems, as we shall see. But it is much loved. Everyone I met who had worked there seemed to have feelings of affection for it. A good mine is like a good regiment; it is the sum of its people but it has its own *élan* which it imparts to them. Herman Lombard, then manager of West Drie, told me:

I have been back here as manager two-and-a-half years. Earlier I was Assistant Manager for three-and-a-half years. And at the beginning of my career I did my four years' training here. I would do it all over again – the best years of my life. It is the people who have established the culture on this mine. Whatever they want to do, they do well. They are real doers.

No mine has ever produced as much gold as West Driefontein. From February 1952 when it first started milling, up to the point when it was amalgamated financially with East Driefontein (though the two are still run as separate mines), it produced over 1,500 tonnes of gold – nearly two per cent of all the gold ever accumulated by mankind. Driefontein Consolidated recovers eighty tons of gold a year, enough to make twenty million wedding rings. Its shares are quoted in Zurich, Basle, Geneva and Brussels, as well as in London, Paris and Johannesburg, and in over-counter trading in New York and West Germany. Of its shares (in 1984), Consolidated Gold Fields had 10.8 per cent direct and 30.4 per cent through GFSA; Anglo-American Gold Investment Company 11.5 per cent; French investors 5.2 per cent; and Barclays National Nominees Limited 5.4 per cent. All other shareholdings were below five per cent. The profits are enormous, though much goes in tax. Thus in 1984 the net profit before tax was R849 million, of which the State's Share and of Tax took R481 million, that is, eight per cent of all corporate taxes in South Africa. Of the rest, R280 million was distributed in dividends, while R112 million was spent on developing the mine, most of it going on eight different shafts. Mines of this quality can always expect to make profits provided they are sensibly developed. Peter Fells of Gold Fields in London, executive director with special responsibility for economic forecasting, put it like this to me:

> Gold is in a way like any other industry in that you hope that you find the deposits and nobody else does. If you find a really good deposit, however, the cost advantage will provide an economic rent for a large number of years. By contrast, in manufacturing industry your initial cost-advantage is very rapidly eroded. The two Driefontein mines are the lowest-cost major gold-mines in the world, and the combined operation has a life of forty to sixty years. Their competitive position is extraordinarily strong, and it doesn't matter if there is over-supply. So unless someone else finds deposits of that richness, they will enjoy an economic rent for the whole of their working life.

West Driefontein is by far the older mine of the two. Before the consolidation it was calculated to have a life of only ten more years. East Driefontein had an estimated life of seventy years. By taking in new mining areas to the north, and then dividing the total mining area diagonally, each of the two mines was given a life-span of about half a century. The performance of the two is broadly similar, though West Driefontein is bigger and employs more people: an average of 15,369 as opposed to 12,670 in the East mine. In 1985 the production costs of the two mines averaged $102 per ounce of gold, the lowest in the entire industry, which means these mines would still make a profit even if gold fell to a third of its 1985 price. But expressed in working profit per ton milled, which is what gold-miners prefer, West Drie's performance is better; R138.33 per ton milled, against the East's R132.50.

No one who has not visited a big, deep-level gold-mine on the Rand can conceive of the scale and complexity of the operation. I quote from my diary:

ABOVE The West Driefontein
No. 5 concrete headgear in
1952. From the top visitors
enjoy a panoramic view of
neighbouring mines.

LEFT In 1952 the first bar of
gold at West Driefontein was
poured by the chairman of
Consolidated Gold Fields,
Robert Annan, who with his
predecessor John Agnew was
responsible for authorising
development of the West Wits
Line.

The surface installations at West Driefontein.

Monday, 18 February 1985. Up at 6 a.m. Around 7.30 reached head office of West Drie mine. Lawn, brilliant flowers, three fountains – the symbol of the mine. Herman Lombard gave me breakfast. He told me: 'We are not the highest grade mine at the moment – Kloof is. On the western side of the mine there was the real gold-mine of paradise. It was chiefly Carbon Leader, which is the most consistently rich of the reefs, though VCR was very rich in places. There, some of the grades were up to 100 grammes a ton. But that was ten to twenty years ago. Unfortunately, most of the top quality gold was mined when gold was only $35 an ounce. That was the problem in those days – low gold prices. You had to mine the high grade ore to make a profit as your costs were mounting while the price remained low.' Lombard took me to the model room, where he gave me a little talk about the topography of the mine and then showed me the big model which, being three-dimensional, gave me an idea of the sheer size of the mine, with scores of miles of tunnels and a surface area of thirty square miles. They extended the tunnels by nearly 30,000 metres last year. The mine's Chief Geologist gave me a history of the mine and explained the formation of reefs. Showed me bits of each reef. They are conglomerates. Pyrite glitters in the dark rock, with its white pebbles. The gold is too fine to be visible.

Lombard then took me to the management's changing-room. I stripped to the skin and dressed in pants, vest, white overalls, two pairs of

Two shafts, Nos 7 and 8, under construction at Driefontein which will come into production during the 1990s.

socks, boots, hard hat, and on top a jacket of leather and cotton. Drove to Number Six shaft. The headgear are of two types. The older ones are of steel. Then, when steel became very expensive in the 1970s, they were made of concrete. This involves less maintenance, painting and so on, and it makes it possible to have lifts up to the top. This is important, as some of the headgears are very tall: the Number Four at the President Steyn mine is ninety-eight metres, the same height as Big Ben. But, with the collapse of the steel price, steel is coming back in again.

Near the headgear I was given a belt and a lamp, equipped with a twelve-hour battery, in what the sign called 'Europeans' Lamp Room'. Lamp is clipped to hard hat. The batteries are recharged daily. Went to shaft. The lift or cage takes twenty men on each deck. Deep-level Rand cages carry up to 150 men at an average speed of forty-five kilometres an hour, up to two-and-a-half kilometres. The hoists lift twenty-three-tonne loads of rock. Enormous motors are needed to develop the power. Number One shaft at Deelkraal has a motor of 8250Kw, equal to 120 motor cars pulling together.

Getting down the whole shift takes two to three hours. The morning or main shift goes on at 4 a.m. or 5 a.m. They work about six hours at the face, but the length of the shift may be up to ten hours because of travelling and waiting time. In a really large mine, such as East Rand Proprietary, the third largest producer, the combined length of the tunnels stretches to over 1,500 kilometres, the distance between Johannesburg and Cape Town. Of course the amount of travelling time taken depends on the number of shafts, and rich mines like West Driefontein have more shafts.

The maximum depth of the main shaft is 2,000 metres. You then transfer laterally to a sub-shaft, a further 1,000–1,500 metres down, and if necessary to a tertiary shaft below 3,000–4,000 metres. At Western

A drilling team at the face in a GFSA gold mine.

Deep Levels the Number Two tertiary shaft goes down to 4,265 metres. That was the record in 1982, but they have since gone deeper. I went down to sixteen levels at about 1,400 metres (I think). The cage fell at eight to twelve metres a second and seemed to go down for a very long time. Some lifts are faster. One goes up to 65.8 kilometres an hour, twice as fast as the quickest lift in Tokyo's 'Sunshine 60' building. I felt no sense of claustrophobia. After the sub-shaft lift we walked along rails to a place where we were given drinks of orange squash. Very hot. The miners wear a sort of rag-towel round their necks for wiping the sweat off their faces. Refrigeration cools the water they use to suppress the dust from the rock-drill machines and blasting, bringing the temperature down from 31.5 degrees centigrade to 28 degrees. This reduction of three-and-a-half degrees makes a big difference. They have five huge fans and seven boosters which provide ventilation by forcing through 230 cubic metres of fresh air per second through thirteen kilometres of shafts and nineteen kilometres of tunnels.

As we got nearer to the actual stope, life became rougher. They spend the morning drilling holes for blasting – 10,000 a day at this mine. At about 1 p.m. they put in the following: three sticks of dynamite, or similar; then a stick of tamping to concentrate the blast into the rock; then a delay fuse made of cotton, which adds two-and-a-half hours to the delay-time after ignition. They light this, then push off. In mid-afternoon, it explodes. They allow for dust to settle and noxious fumes to disperse, fire inspection, etc. Then at 9 p.m. the night shift or 'cleaning shift' comes on and cleans up, getting the ore down the stopes. They push it down with big steel scrapers. According to Lombard, they use an enormous amount of explosives: two tons of dynamite a day, or more.

As in most mines, the ore initially goes down, not up. We scrambled down the stope on our bottoms. Some use little sledges to slide down, but they did not have any here. The gap was only about two-and-a-half feet high, secured by props made of alternating layers of wood and concrete blocks, or just columns of timber. Very unpleasant. We reached the bottom, where men were drilling for new blasts. Then we moved horizontally for a time, and climbed back up to the level where we started. We were very hot and dirty. You must wear gloves at or near the face. Hard hats are essential as I found I was constantly bumping my head. The tunnels are secured by driving steel pins into the rock a metre or more, then stretching wire mesh over it and lacing everything up by hawsers crossed from one pin to the next. Sometimes they spray concrete over it all as well, for permanence. There are sixty-six stopes active at present in this mine. When the ore gets to the bottom of a stope, trucks take it and drop it through holes, or ore-passes, where it goes to the main shaft to be lifted to the surface. Shaft-lifting capacity is one of the keys to successful gold-mining. A big main shaft has double-drum winders of five or six metres in diameter and can hoist up to 400,000 tons of rock a month. But these giants are very expensive. A headgear can take 4,700 cubic metres of concrete, 400 tons of reinforced steel and 800 tons of structured steel. Then the shaft itself requires tens of thousands of cubic metres of concrete for reinforcing. I was told that the Western Deep Levels Number One shaft project, due for completion in 1992, was costed at its beginning at R700 million, an enormous mining development by world standards.

Then to reduction plant. This had high-grade section (mainly Carbon Leader) and lower grade (mainly VCR). Intense security. When the locked door is opened there is a shrill screaming noise. I was shown an illuminated process-flow model which explains the process whereby the gold bars are made very clearly and saves a scramble all over the huge plant. The climax is when the gold concentrate goes into the arc-furnace. Three huge, white-heated prongs are lowered into the crucible, then lifted out when the right temperature is reached, and the crucible tips over. The fiery stream which tumbles out over a succession of seven cups is mere slag – the gold is in darker streams which come out right at the end into the two moulds. These are then taken on a fork-lift truck, seized by men in huge pincers and turned out, then taken by fork-lift again to a water-spray tray where they are cleaned by intense jets. At this point they begin to look like gold bars. These primitive gold bricks are eighty-eight per cent gold, ten per cent silver and two per cent base metals. They then go to the Rand Refinery where they are refined to 999 parts per thousand gold for industry and the arts, and 996 parts per thousand for monetary purposes. ... Finally to the top of Number Five shaft for a spectacular view over the whole Driefontein area and beyond. One notices how inconspicuous the waste dumps are. There are over 400 dumps in the Transvaal and Orange Free State and in the old part of the Rand they are enormous and ugly. At Randfontein Estates there is the biggest man-made sand dump in the world, containing forty-two million tons of milled rock. One dump, at Consolidated Main Reef, is 111 metres high and if they were to put all the dumps on top of each other they would be higher than Kilimanjaro (5,964 metres). But here, in the newer part, the dumps are lower, flatter, of a more geometrical shape. I noticed, too, that the milling and reduction plants, being more modern than those further east, are much smaller (though more efficient and with a bigger throughput). ... It is a chilling thought that, under this enormous area, practically the size of an English county, hundreds of thousands of men work every day, miles deep. Beneath my feet in this one mine there are 900 kilometres of underground tunnels, as far as from Birmingham to Geneva, and this shaft is the top of a system which goes down to 4,100 metres.

In fact West Driefontein has something of a reputation both for disasters and for surviving them. In this part of the Rand, the dolomitic structures over the reefs produce two quite separate hazards. The dolomite is full of fissures and may become like a kind of sponge. The water seeps into it and, being slightly acidic – like a mild acid rain – it eats colossal holes out of the rock, which then fill with more water. If, for some reason, the water drains away, sinkholes develop under the surface cover. When heavy rains wash away the pebble crust, the sinkhole suddenly opens up and engulfs

Pouring molten gold to form bars at a Group mine.

everything on top of it. At Doornfontein there is a sinkhole thousands of years old; people picnic there. But you can never be sure when a sinkhole is quite stable.

On 12 December 1962, without any warning, the seven-storey building, eighty-five feet high, which contained the West Driefontein milling plant, vanished into an enormous sinkhole beneath it, taking with it twenty-nine men. Only heroic efforts by rescue teams prevented the loss of life from being much greater. Equally impressive were the efforts to maintain output, as Peter van Rensburg explained to me:

> Management arranged for the railways to lay on trains. The mine kept working. Ore was taken by train to Libanon and Venterspost and by road to Doornfontein. Steel balls were added to the mills, and they milled on Sundays. By the end of the month the mine was once again producing at its full capacity. There was tremendous cooperation between our mines and outstanding assistance from the railways' people. Head office ensured fair shares and financial adjustments. The most significant thing was that West Driefontein never dropped its deliveries, the true spirit of the mine. There were also prodigies of effort to find a new site and design a completely new plant. It was put on an area with Pretoria Series rocks which are impervious to water – so the dolomites beneath don't matter. By 12 December 1963, a year later, the new plant was running.

The same year the Far West Rand Dolomitic Water Association was formed. All the mining companies took part. It handles the problems arising from de-watering of the dolomites, which include claims from farmers who lose their water. It owns a lot of ground bought up as part of settlement for damage suits. The state has given permission to mines to de-water because gold recovery is much more valuable to the country than agricultural production of that area.

Throughout the 1960s there was something akin to panic over the dolomitic water problem and the risk of flooding. Most of West Driefontein, Blyvooruitzicht and Western Deep Levels are below a vast underground lake called the Oberholzer Compartment. It cost these three mines over R200 million in pumping costs in the year 1968. From 1955 West Driefontein never pumped out less than seven million gallons a day. In 1962–64 it had to pump out over thirty million gallons a day, and it soon had more pumps than anywhere else in Africa, capable of handling up to sixty-three million gallons daily. As van Rensburg put it: 'West Driefontein has to get to the surface three times the tonnage of water to rock – you could call it more of a water-mine than a gold-mine.' At one time enough water was pumped out of West Driefontein each day to supply a quarter of the needs of six million people living in the Witwatersrand, Pretoria, Southern Transvaal and the Orange Free State.

The mine took other precautions, besides building up its pumping capacity. It fitted a series of watertight doors to protect the pumping shafts and the pump crews up to level 16. In effect this provided a sort of reservoir or water storage capacity of 1,000 million gallons at the bottom of the mine, and in case of flooding, until this filled up, the pumps could still work.

Unfortunately, 1,000 million gallons is nothing compared to the capacity of the big dolomite oceans. In 1966 West Driefontein began to move east, sinking its Number Four shaft to mine the VCR under a water area known as the Bank Compartment. This stretches over sixty square miles and contains at least 100,000 million gallons. The work involved driving through a huge fault, known as Big Boy, which extends up to the base of the dolomite where the water-bearing zone begins. The Bank Compartment had never given much trouble to Libanon and Venterspost, which also mined under it; these mines had never pumped more than three million and eleven million gallons a day respectively. West Driefontein's first two years in the Bank Compartment brought no problems, until Big Boy intervened. Work down to 10 and 12 levels went ahead, both being connected by a 12,000-foot tunnel to Number Three shaft.

Then, without warning, on Saturday, 26 October 1968, a fissure opened between 4 and 6 levels at Number Four shaft. For the next three weeks a desperate struggle took place to save the entire mine. It has been recorded by the mining historian A.P.Cartwright, who got eye-witness accounts of those involved immediately afterwards, in one of the best books ever written about the Rand, *Ordeal by Water*. The cause of the fissure is not known, the most generally accepted theory being that a slight earth-tremor on Friday night started a rock-movement. The exact place and size of the fissure is unknown even now. It may have been as small as 140 square inches: a crack one inch wide and 140 inches long at that depth can let eighty-two million gallons a day into the mine.

The first the surface management knew about the crack was when it received a message, at 9.40 a.m., on a scratch pad, reading: '6/46B 4W

LEFT On 26 October 1968 a flood at West Driefontein demonstrated the terrible power of water – the rush of water down the shafts, sucked the air out of some workings to create new vacuums. Fierce winds of up to 130 mph followed as the vacuums were filled. There were no casualties, but water continued to flood the No. 4 shaft at up to 100 million gallons a day. On 18 November plugs were installed just in time to contain the water which would otherwise have submerged the pumps and closed the mine.

Broken Open 6″ into +′. Translated this meant: 'Stope 46B on 6 level West of Number Four shaft has broken open and water is running six inches deep into the shaft.' At first the full seriousness of the flood was not grasped; then, when those below calculated that the inrush of water was at the rate of over ninety million gallons a day, and the cage in Number Four shaft began to yo-yo, the decision was taken to get the men out. They were told to make their way to Number Three shaft, trekking along the tunnels. Some had to climb up to 12 level, then start walking west.

All the men got out except two: Alberto Noife and Luiz Sandela, both Shangaans from Portuguese East Africa, who were trapped working in the tiny pump compartment in Number Four shaft at level 13 3/4 and between 12 and 14 levels. Two senior men, W.C. Theron, a European shaft timberman, and A.Vascoe, African Boss Boy of Number Four shaft, volunteered to go down in the violently swinging lift-cage to get them out.

Vascoe later recounted:

> On the surface I had not been afraid, but I must admit that when I saw the amount of water, and the noise it was making, I was afraid, but I felt something must be done to save the pump-boys. When the cage stopped Mr Theron was informed through the walkie-talkie that we were at 13 3/4 level. We could not see the station as the water was pouring past the cage. I left the cage and went to find the pump-boys. They had removed their boots and hard hats and they were very scared. I made them put their boots and hard hats on and we returned to the station. They were afraid to climb through the water into the cage, so I climbed back into the cage, collected a rope and climbed back into the station again. I tied the rope to the first Bantu and while Mr Theron held the rope I made him climb through the water into the cage. I did the same for the second Bantu, and I then returned to the cage and we were pulled to the surface.

This laconic account conceals the true heroism of the rescue. Vascoe and Theron were each awarded the Chamber of Mines Bronze Medal, the 'miner's VC'.

So all were saved, and once the men were out the next priority was to contain the flow and prevent it getting along 10 and 12 levels into Number Three shaft and so ruining the heart of the mine. The rush of water was not the only hazard. Its movements, pouring down the shafts, drawing the air after it, created near-vacuums and thus gales of fierce wind, as a result of the venturi effect. These gusts have been measured at 133 mph and in narrow parts are so powerful that nothing can stand in their way. Some of the men engaged in work to save the mine had to be pulled to safety, while lying flat on their stomachs, by power-operated chains.

Stopping the flow required two measures. Plugs had to be inserted at 10 and 12 levels, to save Number Three shaft. And, in the meantime, sufficient pumping capacity had to be generated to contain the flow and stop the water-level from drowning the pumps and putting them out of action. At one point the water-level came within a few inches of key pumps before it started to fall. But the pumping crews managed to keep all the pumps working at top capacity throughout the emergency and, despite the appalling conditions, prevented a single breakdown. More, they installed new pumping capacity, which would normally have taken two months, in a mere three to four days. In addition, two adjoining mines, Western Deep Levels and Blyvooruitzicht, drilled through to the mine to add ten million gallons a day pumping capacity. Even so, it was touch and go for a fortnight.

For the plugging, Gold Fields' former associate and long-time partner, Cementation, was called in. Plugs are walls of solid concrete, from ten to one hundred feet long. To make an emergency plug you have to lay pipes and put in the concrete around them. Water continues to pour through the pipes while the concrete is setting. When it has set, the valves on the pipes are shut and the seal is complete. Then the barrier is hastily made secure by pumping cement through special pipes to the 'wet' end, filling all the pipes with cement, and building further barriers at the 'dry' end. By the finish, a solid wall of cement hundreds of feet thick blocks the tunnel. It sounds simple but it is a complicated and painstaking process because the walls of the rock have to be brushed absolutely clean. Essentially, the mine was saved by the expertise and energy of the Cementation people, and the men of West Driefontein, as well as the skill and heroism of the pump crews.

It is worth adding here that Cementation is not just an emergency service. It is essential to normal deep-level mining in the Rand, and to mining operations of many other kinds. The nub of Albert François's invention is that he managed to demonstrate that you could inject cement or grout into cavities even under water, provided you use enough pressure. As Cartwright says: 'Improved and refined, the process he devised has done more to assist mining engineers in all parts of the world than any other single invention this century.' The Cementation Company (Africa) Limited is the South African branch formed to work François's patents. It has a team of forty men permanently employed at West Driefontein. All big mines have Cementation ranges, or delivery pipes, which run down the shafts and carry grout (sand and cement) almost anywhere in the mine. The only problem is to prevent blockages in the pipes, which tend to occur about once every ten days. But so long as the pipes are free, fast-setting cement is on tap everywhere it is needed: you press a nozzle on the pumps and it flows out to fill any cavity.

Cement on tap is only one of the ways in which high technology is being applied to make the mines more efficient, comfortable and safe. The deeper you go, the more mechanisation and technology you require to maintain productive efficiency. As the former Gold Fields' Consulting Engineer, Dr Taute, put it to me:

In the old days (1940s) the labourers were simple people. The mines were labour-intensive – tramming gang, haulage gang, stoping gang and so on. They took all day to load the ore by hand. But we still achieved thirty tons of ore per man per month, and this included all auxilliary sweeping, track maintenance, ventilation, etc. At Libanon we got over fifty-four tons per man-month. With all today's mechanisation we are back to about thirty tons. This is mainly due to depth and heat. The distance travelled underground reduces the length of the operating shift. Indeed, without mechanisation you simply couldn't run these deep-level mines.

Some of the new methods involve very simple ideas. You wonder why no one ever thought of them before. Take heat. The virgin rock temperature at the bottom of the mine is about 60 degrees centigrade, 140 fahrenheit. Heat increases in accordance with the Geothermal Temperature Gradient. On the Witwatersrand, for every 212 feet you go down, it rises by one degree fahrenheit. In the Orange Free State it's one degree for every 160 feet. The West Wits rock is over 2,500 million years old and sterile. When you make an excavation, the rock exudes heat. After World War Two, the fashion was to refrigerate the air, so cool air was put in with vast fans. The air going down picks up heat, and by the time it gets to the works it is quite hot. About five or six years ago, the Chamber of Mines' research organisation began to chill water, sending it down a pipe insulated with asbestos lagging. In accordance with the strict anti-dust regulations, you have to spray water in the workings to lay the dust anyway. After every blast, the stope must be watered down. When you use ice-cold water to wash your stope, this cools down the working area – and iced-water is much more efficient than cooled air. The chaps now sometimes have to put their coats on and this is one hell of a breakthrough.

The man who devised the new system was Dr Austin Whillier, Director of the Environmental Engineering Laboratory at the Chamber of Mines. A sub-tropical city like Durban has a midsummer 'wet-bulb' temperature (a measure of heat and humidity) of 27.5 degrees centigrade. Above this, men must be acclimatised before doing hard manual work. This means four to five days in surface acclimatization-rooms, taking vitamin C tablets. More than 300,000 black mine-workers have to be heat-adapted each year. At 2,500 metres depth, the rock temperatures range up to 31 degrees centigrade. By lowering it 2 degrees centigrade, production can be increased by forty per cent. Under the old air-cooling system on one of our mines, some fifty per cent of wet-bulb temperature observations at the deep levels exceeded 30.5 degrees centigrade. Then they began to pump down iced-water: at West Driefontein, for example, 35,000 tons of iced-water is pumped through the system daily. A year after the new system was introduced, there were no observations of wet-bulb temperatures of 30.5 degrees centigrade and less than twenty per cent over 28.5 degrees centigrade. Cutting down those 2 degrees makes a huge difference; and, by fitting taps, the water can be drunk too.

Though all the major firms in the industry have contributed to technical advances, Gold Fields is justifiably proud of its record in this area. Harry Cross, formerly in charge of Metallurgical Developments, put it this way:

Gold Fields has always been on top technologically because we have investigated the fundamental phenomena governing the recovery of minerals – for instance, colloidal substances in ore. We had a lot of firsts in recovery of tin, zinc, fluorspar – our new processes being the by-product of research on theoretical considerations. Gold Fields has always been good at recovery. This is why not a single one of its old dumps have

OPPOSITE ABOVE An early view of Kloof, which dates from 1968, showing the No. 1 shaft.

OPPOSITE BELOW Mineworkers at Kloof, 6,000 ft below surface, excavating a 50-ft diameter chamber to house the head frame of a sub vertical shaft.

A shaft-sinking platform carrying men and equipment during excavation of the No. 4 shaft at Kloof.

been bid for – Gold Fields got most of it out in the first place. Our Wanderer miner in Zimbabwe was the first flotation gold mine in Africa another success story. At Motapa near Bulawayo, we eliminated arsenic by erecting a roaster plant, in which the arsenic was driven off. That was another Gold Fields' first. In 1951–52 we were also pioneers in the grassing of dumps. This has largely eliminated the terrible dust which used to be the bane of Johannesburg. We did the experimental work on grassing and it was then taken over by the Chamber of Mines. This technology of re-vegetation, which we developed, was later exported to Australia to grass over exhausted mineral sands' workings.

We were particularly proud of the first 6,000-foot shaft at Driefontein. It was less than three centimetres out of vertical. We put the first giant mill in at the Venterspost mine and Gold Fields pioneered the first electronic controls. Today, Gold Fields is still in front technologically. There is nothing fly-by-night about its innovations – all based on sound technological principles. Our technical men strive to be the top technical men, not managers.

Dr Taute gave me further details of innovations in mining engineering in Gold Fields' operations:

We used electric motors to replace the men who opened and shut the double ventilation doors. There are so many doors in a large mine you can easily save fifty men. I wrote a paper in 1946 proving you can profitably spend up to £2,000 on investment for every man saved. We used tungsten-tipped steel for drilling blast-holes. Previous bits only drilled one hole. We analysed how best you can improve flow of rock from stope to surface, and flow of men and materials from surface to face. I specialised in shafts, underground stations, track layouts and so on. Designing the Kloof mine presented an opportunity to apply in practice what had been worked out in theory. Its main shaft was then the largest in the world – 31 feet 6 inches in diameter, 6,700 feet deep, hoisting 450,000 tons per month and providing facilities for men, material and ventilation too. It had ten conveyances, five hoists, and the skip was a totally different concept, bottom discharge instead of overturning in the headgear. We also designed the underground stations so that they were correctly orientated to suit the dip of the strata, facilitated optimum flow of men, material and rock and were large and cool.

A boom drill rig at the bottom of the Kloof No. 4 shaft. A good example of how mechanisation of traditional methods is helping to speed up shaft sinking on Group mines.

The recent changes in development have been enormous. The drill platform is pushed into position mechanically. One chap operates the mechanical loader – used to be a gang of twenty. Tunnels used to be as small as possible. Now you walk upright, in a flow of cool air. So conditions are actually better than in many of the older, cooler, shallower mines. Gold Fields was in the forefront of all these changes. We were the first to introduce a rock-borer in hard rock. It didn't work – a lot of our ideas didn't work, at first. But it did in the end, and rock-boring has come to stay: it's not so good for horizontal tunnels but for inclined ore-passes at fifty-five to sixty degrees it works exceptionally well.

Our next idea was the manless stope. It is in the stopes that nearly all the rockfalls and accidents occur. The trouble is that the stope is usually very narrow, and apart from using the mechanical scraper you still need labour to this day. The machinery we devised – the stope borer – worked but it was too cumbersome and expensive. The concept was sound but execution poor. The operators lacked enthusiasm to work it. Whenever you introduce something new into a mine it either takes off like a grass-fire or you have to fan it – and if you constantly have to fan it, it's no good. All the same, I am sure we will get a manless stope one day.

During the 1950s we broke the world's shaft-sinking record. We were the first to introduce a cactus-grab. An engineer watched a child grabbing sweets in a bowl and imitated it mechanically – loading two tons of rock into the sinking kibble at one go. This revolutionised shaft-sinking. When I was underground manager at Doornfontein in 1956 we broke the world record for hard-rock tunnelling – 1,905 feet in twenty-six days and 2,170 feet in thirty days, using standard equipment. The training did it, everyone knowing exactly what to do. A Free State mine eventually did better, but only by using equipment especially designed for this purpose.

The great thing about gold-mining is that new ideas are shared immediately for the benefit of the industry as a whole. The Consulting Engineers Division Annual Report, founded in 1974–75, gives details of all Group technical innovations during the year. We can do this because there is no competition in selling gold. With the base metals industry it is quite different. Producers are competing in a limited market and everything is a secret, cloak-and-dagger stuff – the longer your competitors struggle to emulate your success, the higher their costs and the more likely they are to go out of business. In gold there seems to be no limit to the market so there is absolute freedom in sharing technical innovation. That is why the South African gold industry is so efficient. It is the fact that everything is shared which makes it a great industry.

A safe industry too, in all the circumstances: for in deep-level mining over a mile below the surface you cannot separate ideas and new technology from safety. The Rand once had an appalling reputation for losing miners through accident and disease. A survey made by nine mine doctors from November 1902 to April 1903 showed that, during this six-month period, 1,541 out of 50,000 black miners died from accident or disease, an annual death rate of 61.6 per 1,000. The overwhelming majority, 54.5 per 1,000,

ABOVE Mineworkers installing a roof support.

LEFT East Driefontein dates from 1972. A cactus grab filling kibbles with broken rock while sinking the north shaft.

BELOW LEFT Mineworkers drilling a stope at Doornfontein which dates from 1954.

ABOVE The No. 3 shaft at Doornfontein.

RIGHT Travelling by chair lift at Doornfontein.

died from sickness. At the Simmer and Jack mine, of 206 recruits arriving on 26 February 1903, 116 were dead by June, equivalent to the death-rate during the Great Plague of London. The main killers were pneumonia, meningitis, enteric fever and dysentery. Mine accidents killed a much smaller proportion but as the depth increased it was rising: 4.7 per 1,000 per year in 1907, 5.26 in 1909.

The mining industry tackled the problem of accidents directly in 1913 by creating the Prevention of Accidents Committee, now known as the Mine Safety Division of the Chamber of Mines. By the end of the First World War it had reduced the fatality rate to 2.81 per thousand a year. From 1922 it provided training through the South African Red Cross, and from that date over 3 million black mine workers have received first-aid certificates; today about 100,000 qualify or renew certificates annually and 20,000 take part in mine first-aid competitions. By the first half of the 1970s, the death-rate had been reduced to 1.25 per thousand a year.

By this point, however, it was recognised in the world mining industry that isolating accidents from other aspects of mining was not the best approach. As Dr. S.Lancer of the Operational Research Department of the British Iron and Steel Research Associations put it in 1973:

> What goes under the name of accident prevention is, in the main, injury prevention ... parodoxical as it may sound, this exclusive concentration on injury-causing accidents is also the chief obstacle to further big steps towards their prevention.

Accidents were once measured by cause and consequence. The missing element was control. In 1976 the South African gold-mining industry adopted the principle, already being applied in the United States, of loss-control management. This implies the control not only of accidents which cause personal injury but of all accidents which cause damage to material and equipment and lead to losses, because they all spring from similar causes. Loss control has been officially defined as 'The application of professional management techniques and skills to those programmes, arrangements and activities specifically intended to minimise the losses resulting from undesired events arising from non-speculative risks of business.' It is concerned with pin-pointing the risks, analysing them, and then devising management controls to ensure that all work is carried out to clearly defined standards. This brings accident prevention automatically right to the heart of management and of all training. Hence increased productivity and increased safety go hand in hand.

In 1978 the South African mining industry and the International Loss Control Institute in Atlanta worked out the International Mine Safety Rating Programme. This provides for internal and external measurement of control of accidents to persons, damage to property and 'unplanned interruption of operations'. The rating system has twenty-two elements, ranging from training and equipment to reference libraries, and it awards from one to five stars: for five stars a mine must achieve over ninety per cent compliance with all twenty-two elements and twenty-five per cent below average for fatalities and reportable injuries for its class of mine.

The effect of this new approach has been marked. In April 1984 West Driefontein, with a five-star rating, had recorded one million fatality-free shifts; East Driefontein, also raised to five-star status, recorded a million fatality-free shifts two months later. Gold Fields' mines stand high in the ratings system which, of South Africa's seventy-odd mines, classifies (on average) about twenty as five-star, twenty as four-star and ten as three-star. Granted the exceptional characteristics of South African mining – the unparalleled depths of operations in the gold-mining industry and the high proportion of miners recruited as unskilled labour from rural areas – its safety record in the early 1980s was impressive. In coal-mining, for instance, during the first six months of 1983, fatalities were 0.53 per thousand per annum, as against 0.6 in the United States and 0.56 in West Germany. In the gold-mining industry, working at depths up to four kilometres below the surface, the fatality figure was higher: 1.25 per 1,000. But if pressure-bursts, the direct consequence of mining at great depths, were excluded, the figure dropped to 0.95, almost as low as the corresponding metal mining figure in the US of 0.93 per 1,000. (This was written before the Kinross disaster.)

One reason why, granted the extreme conditions, safety standards in South African gold-mining are now so high is that these vast mines are run in a highly regimental manner. Discipline is as strict and as automatic as in the Brigade of Guards. Indeed, it has to be, for to mine at such depths is to

Deelkraal started working in 1980. These three photographs show the No. 1 shaft collar under construction, installation of the semi autogenous mill and a general view of the mine.

fight an unrelenting war against the elements in which death or injury is to be expected any second and obedience to orders must be instant and unthinking. For each stope panel there are seven unskilled labourers. Then four drillers, semi-skilled and one skilled team leader, in charge of three groups of seven (each seven having a semi-skilled team leader). The section leader is capable but black, so not allowed under the law to be classified as 'skilled miner'. Each white skilled mine-worker has two groups of teams. He is called a ganger or stoper. Three gangers report to a shift boss, who is thus in charge of about 150 blacks. Four shift bosses report to the Mine Overseer or Mine Captain. Four Mine Overseers report to the Underground Manager. Three of four of them report to the Assistant Manager. The Mine Manager looks after everyone. Things are very formalised, army-style. To the man in the rank above you, you always say 'Mister' and he calls you 'Joe' in return. Down it's Joe, up it's Mister. This is essential to discipline when things go wrong, as they inevitably do from time to time. (Knowing names is important: Herman Lombard, formerly manager of West Driefontein, told me he knew eighty per cent of the 1,500 white miners in his mine by name; with the 14,200 black miners he reckoned his score at twenty per cent, but then there is a higher turnover among the unskilled men.) You cannot visit a South African gold-mine without being made aware at every step of the military discipline and regimental élan which pervades it – so different to the informality of mines in the US and, still more, Australia. Safety is emphasised everywhere. A sign shows hands: 'Your ten best tools – all irreplaceable'. Most mine accidents affect hands. In the Rand there are now fifty-one hospitals catering for mining injuries, including the 1979 Rand Mutual Hospital which specialises in difficult mining cases. One of these mining hospitals is Gold Fields' Leslie Williams Memorial Hospital. My diary records:

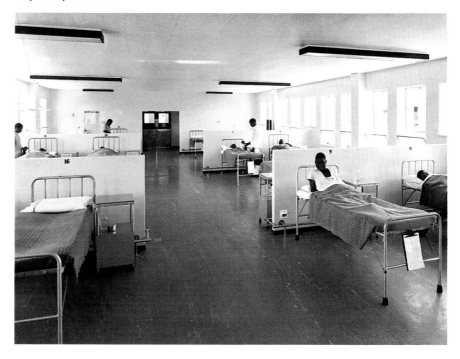

The Leslie Williams Memorial Hospital for miners near Carletonville.

Joined by Dr F. J. Duminy, superintendent of this 550-bed hospital which serves the two Driefontein mines and two others in the group, 44,000 miners in all. He took me on a detailed tour. It treats only males, but they are adding a new wing for wives and children, including visiting wives, who can have inspection and treatment. Of the hospital's cases, fifty per cent are accidents, many to hands. Lots of broken and crushed bones. He showed me a bad case. He has just sewn on a man's thumb. Surgery of the hands is his speciality. He has recently been on a microsurgery course in San Francisco and he showed me an elaborate new microsurgery microscope and TV links used in operations, allowing pupils to watch. There are the latest spectroscopes for internal examinations, and a big surgical operating area, with an elaborate flow-system to keep it sterile – he was able to incorporate this feature in the architect's plan. He is very proud of his hospital and rightly so.

These specialist Rand hospitals not only mitigate the consequences of mine accidents but have cut the disease rate per year among the workforce from over 50 per 1,000 to 1.8; pneumonia is down to 0.7 per 1,000. The hospitals are part of the miniature welfare state run by the Rand mining finance houses. Broadly speaking, all services provided to black miners are free. These include X-rays on first employment, at six-month intervals and on discharge; hospital treatment; haircuts, dentistry, temporary grazing areas for cattle and special bus services to homelands; and, above all, food and accommodation. At West Driefontein I noted in my diary:

Safety is a priority on Group mines.

Visited big bachelor complex for black miners. Run by European manager and assistant, and two black assistants. Fairly new. Men are here on contracts which last from forty-five to fifty-two weeks. If their families live reasonably near, the bus service gets them home for weekend visits. There are units of eighteen men with cubicles, plus dining area and hotplate for cooking. They elect a room representative, responsible for cleaning. Manager says they seem to prefer older-style barracks, nearer shaft-head, so involving less travel. Went round dining hall, then kitchens. Purchasing on vast scale for entire group, and cooking on pretty big scale in each kitchen enables them to keep cost of feeding to R1.20 per man per day for well-balanced nutritious meals. They will now eat fish (once taboo) and sausages (ditto). Married rations for husband (wife draws three-quarters, child half each): meat 2 kilos a week; fish $\frac{1}{2}$ kilo a week; chicken 1 a week; bread 1 loaf a day; fat $\frac{1}{2}$ kilo a week; beans 2 kilos a month; maize meal 10 kilos a month; mealies 2 kilos a month; mealie rice 2 kilos a month; samp 2 kilos a month; barley $\frac{1}{2}$ kilo a month; sugar 2.5 kilos a month; salt 125 grammes a month; instant coffee 125 grammes a month, and tea and cocoa likewise; noodles $\frac{1}{2}$ kilo a month; vegetables 6 kilos a week; cooking oil 1 bottle a month; fruit as available. All this is free.

Taken round beer-garden, open every evening when all sorts of liquors can be bought in bar-room controlled by a Mr Trollope. Outside tables and chairs are of multicoloured plastic. Lawns will not survive so many feet, so concreted over. Canteen serves food. They tend to drink

Hostels for mineworkers at Kloof with a sports ground in the background.

communally, each member of group buying jug or bucket of beer in turn. Nearby is rather grand accommodation for visiting chiefs and dignitaries, with individual bedrooms, bathrooms, lawns, kitchens, TV, etc. – sort of hotel. Also round, kraal-type huts where miners can have wife staying for visits; but only one at a time – I was told they usually picked the youngest, prettiest wife. This privilege is highly valued. I was then driven round married quarters. They are building more, but within the limits of the government's policy which will only allow them to house three per cent of their blacks there. Management would like to raise this possibly to ten per cent, but not more. Married quarters have nice gardens. But mistake made in not putting in garages. So many blacks now have cars that this is a problem.

The recruitment, training and management of large numbers of black miners have from the start been among the most difficult tasks of the South African gold-mining industry; and it is now undoubtedly its principal problem. Because of the central importance of gold-mining to the nation's economy, government has always closely regulated its activities, especially its handling of labour. In the years after the Boer War, the government imported Chinese labour to help work the mines; then, after a Liberal ministry took power in London in 1905, it reversed the policy. The Chinese phase had beneficial effects, inculcating higher discipline and method in the black labour force. Important, from the point of view of the mining industry, are the Scheduled Persons laws (based on earlier legislation) which limit certain categories of jobs to whites only.

The mining industry is strongly opposed to apartheid, as is South African industry as a whole. The system is incompatible with a free market in labour, the very essence of capitalism; indeed, with its concept of immutable, hereditary status (based on race), apartheid has some of the characteristics of feudalism, and it is the nature of capitalism, especially the large-scale capitalism found in the mining industry, to erode and finally destroy feudalism. By the mid-1980s this was apparent even to the Nationalist government, which began to repeal some of the apartheid laws and ceased to enforce others. In industry the system is clearly breaking down and it remains effective chiefly in areas where powerful vested interests – such as the white mining unions – stand to benefit from its enforcement.

You will now meet very few people in the higher ranks of the mining industry who favour any sort of barriers to black advancement. The manager of one Gold Fields' mine told me:

In my time I have seen the black employees turn to mining as a career. Before that the black was a subsistence farmer and mining was a side-job. Now subsistence farming doesn't really exist, so mining is his life. Our job is to improve the black man's education and to give him the opportunities to use it. It's not true it is only the government's job to educate the black man. It is ours too. Of course we have always taught them fanakalo [the miner's ligua franca, a mixture of Zulu, English and Afrikaans]. All miners have to learn it. But we have now started basic literacy classes. We must teach them a good technical language like English. Fanakalo is OK underground but it can't accommodate new vocabulary. We are now getting educated blacks joining the mining industry. Of course, people with certificates are not necessarily very good – you have to test them. The only regulation that really hampers us at the moment is the one on Scheduled Persons. But this has a short life in my view. The Chamber of Mines is determined to go ahead, despite the opposition of white unions. The government has taken the easy way out by saying the Chamber and the unions must negotiate. We can handle any white backlash when the law is changed. It won't bring revolution overnight. We have very high standards. I have a few blacks who are ready for promotion underground. But we need to make rapid changes on the engineering side. We have accepted black apprentices, and we have people ready, but they are mainly foreigners from Mozambique. We also have to change the attitude of the blacks themselves. At present they all want to become clerks instead of engineers, and to work in offices instead of advancing up the blue-collar underground hierarchy. The men I'd most like to push forward are those who have worked underground for five or six years – some have done much longer. In effect at present the white unions have the power to veto, and I think government should simply change the law.

Unfortunately, the black/white divide is only one aspect of the extra-ordinarily complex racial and tribal mixture of southern Africa, which is fully reflected in the gold-mining industry. A typical Gold Fields' mine draws labour from twenty tribes or tribal groups as well as foreign countries like Lesotho, Mozambique and Malawi. Many of their black miners come from the independent states (within South Africa) of Transkei, Bophuthat-swana, Venda and Giskei, each of which is a fairly homogenous racial group. Of the population of South Africa proper, 23,772,000 at the end of 1982, the biggest racial group is the Zulu, with 5,421,000, followed by the whites (themselves descended from Dutch, French, British and Germans) with 4,454,000, the Xhosa with 2,685,000 and Coloureds with 2,556,000; in addition there are 780,000 Asians (sixty-five per cent Hindus, twenty-one per cent Muslims, the rest Buddhists and Christians) and nine major black tribal groups ranging from the 2,265,000 North Sotho to the 185,000 Venda. Even the Venda have twenty-seven distinct tribes and the Zulus have more than two hundred.

Mining finance houses and individual mine managers have to take account of this variety. They have to operate not only through modern trade union organisations but through traditional tribal structures. Wage negotiations with unions are conducted at Chamber of Mines' level and do not directly concern managers; but managers have to be acutely aware both of union attitudes and tribal feelings, which are often in conflict. They argue that some tribal groups are much easier to work with than others. One mine manager told me:

> The Shangaan from Mozambique seem to me to have the strongest moral framework. I attribute this to their Roman Catholic background and strict Portuguese notions of discipline, as opposed to British liberal ideas. These people seem to have higher moral standards than most white miners or than other black groups. They are less liable to drink, gamble, steal, kill or desert their wives and children.

Management must exercise diplomacy in making up work-parties. An isolated miner sent to work with a team composed of another tribe may be at risk. All prefer to spend their leisure-time with fellow-tribesmen. It is Gold Fields' policy that men should live together by tribal affinity. Anglo-American prefers them to live together by work-groups. This is more convenient if you want a section out quickly in an emergency but it is less popular with the men.

In general, Gold Fields prefers to reinforce traditional structures while Anglo-American, which takes pride in its 'liberal' image, promotes modern ones. Anglo-American, for instance, promotes blacks with matriculation certificates to senior white-collar positions in hostels. The Gold Fields' people argue that the blacks do not like this. A man may have his certificate and even a degree, but no personal status back in his homeland. Gold Fields uses the *induma* or tribal style of election, on the grounds that most blacks prefer it. They claim, as one of them put it to me, that 'an errant black will confess to the *induna* but not to an official black personnel manager.' It is true that black miners are traditionally inclined in many ways, and not merely in attachment to tribe. Many come from families which have worked for the same house, or even mine, for generations (as, of course, do many white miners).

But the status and outlook of the black miner is changing fast. Since the recovery of the gold price from the early 1970s, wages of black miners have more than tripled in real terms. They have more possessions, are better educated and more widely travelled than ever before. They are also much more likely to belong to a trade union. Between 1979 and the end of 1984, the number of South Africans of all races in registered trade unions more than doubled; by far the largest increase was among blacks. In 1979 there were very few blacks in registered unions; by the end of 1983 the number has passed the half-million mark. The Rand is one of the greatest concentrations of industrial labour on earth; granted this, and granted South Africa's political and legislative framework, which makes many forms of trade union activity illegal, periodic unrest and even violence are probably inevitable. These incidents rarely last long. The half-million black miners value their jobs and wages, which support more than ten million people over a dozen countries and form a key element in the economy of all southern Africa. The likelihood is that neither Gold Fields' traditionalism nor Anglo-American liberalism make much difference in practice – one old hand from Gold Fields claimed to me that 'Anglo is always having trouble despite, or perhaps because of, its liberalism' and my impression is that the contrast in policies has been exaggerated.

Removing slag from a newly poured gold bar.

On one point all senior managers at GFSA are adamant: in dealing with labour, Gold Fields always sticks to what it promises and means exactly what it says. Adriaan Louw, who ran the company up to 1980, insists:

We never make a promise we don't keep. This absolutist attitude started with Carleton Jones. It works both ways. In my time we had a sudden strike at 7 o'clock. I told them that if they were back by eight, all would be forgotten. Those who were not back would be regarded as fired and would have to reapply for their jobs. We stuck to this.

Robin Plumbridge, the present Chairman, endorses this and gave me what he termed 'my basic philosophy' towards the workforce:

Everyone we employ should have the chance to progress to the limit of his possibilities. Our training schemes should facilitate this. A non-discriminatory promotion system based on merit is crucial. This applies not just to race but to the male–female issue. But it's easier said than done. We have to operate within the law, while actively trying to persuade our political masters to make changes. We have to take account of trade union arguments, but legal constraints are far more important. We have to seek opportunities for communication, to get abreast of their attitudes, bosses with workers, whites with blacks. The boss must understand what black workers are thinking – there is a double

Taking a sample for assay.

dichotomy here. He must also understand what an educated black man thinks in relation to what a black worker thinks – an important point. I personally know many of our senior blacks. All our senior people now make these contacts as a matter of course. It acts as a check on information we are getting from lower down the line. Our senior blacks have direct access to senior line management. When things are brewing in the engine-room we get the vibes coming through. You have to have an alternative route to get correct information to and from the workforce. Our reputation with the unions is tough. We are prepared to stand up and be counted. We are prepared to face a strike. The union leaders know exactly where they stand. Colin Fenton and I were for many years the only ones from any mining group who attended the white miners' trade union annual meeting. We have had a conservative image. So be it. We prefer to work with people. We don't advertise in the press what we're doing. There's no glare of publicity which produces posturing, and brings progress to a halt. I hate tokenism. Everything must be on merit.

The problems of running a multi-racial workforce on the gold-mines have to be seen within the wider political perspective of South Africa's racial present, and future. Nothing is more difficult to predict than what is likely to happen in South Africa in ten or twenty years' time. Happily I am not called upon to do so. But it is worth noting that the apocalyptic vision of South Africa's future, which it was fashionable to outline in the late 1950s and early 1960s, was wholly falsified by events. At that time, with the 'wind of change' sweeping over the old black African colonies, it was generally assumed in the West that the white regime in South Africa would not long endure and that its economy would rapidly lose its dynamism under the twin burdens of apartheid and black nationalism. In fact far from declining, the South African economy flourished mightily from 1962 onwards, throughout the boom of the 1960s and early 1970s. When the boom ended in 1973–74, South Africa was the principal beneficiary of the price-revolution in gold which followed. The South African economy, which is often contra-cyclical, went into recession in the second half of 1984 and this, combined with fresh outbreaks of violence, again raised doubts about its future.

In London, Consolidated Gold Fields tends to be both optimistic and pessimistic about its South African investments. The problem as seen from London is more economic. Peter Fells, CGF Senior Economist and a Director, told me:

> We took the view in the 1970s that dominoes were falling – Angola, Mozambique, and so on – so there was a certain amount of panic. My view was that the South Africans had the means to repel any external aggressor. But in the long-term demography is against them, especially with the increasing urbanisation. A more subtle threat and a more near-term one is that, as the white population shrinks relatively, the sheer burden of the social services, education, law and order will lead to higher taxation which is bound to hurt the economy.

Many wise heads in Johannesburg share this sombre long-term view; many do not. South Africa always looks more solid seen from within than from without. In 1960 it was the only modern industrial economy in Africa. A quarter of a century later it is still the only modern industrial economy in Africa and the gap between its economic power and that of all other African states has widened. Of the forty-odd independent black African states, most have proved gruesome failures; only two or three can be said to have prospered. Soviet Russia has lost much of its enthusiasm for pursuing an active policy in southern Africa and baulks at the expense. In South Africa itself some of the more objectionable features of apartheid are being dismantled: the absurd sex-laws, for instance, went in 1985. It is evident that the regime wishes and intends to broaden the racial basis of its policy. As against this, the African National Congress is pursuing a policy of

Deep level drilling to find new gold deposits on the West Wits Line.

OPPOSITE ABOVE Part of the train-loading facility which serves Apex.

OPPOSITE BELOW Coal stockyard at Apex

systematic violence specifically aimed at terrorising moderate blacks prepared to work with the authorities. This strategy has certainly succeeded in destroying the chances of compromise in other parts of the world. So South Africa's immediate future can be seen as a race between the deliberate liberalisation of white supremacy and the deliberate pursuit of a crisis of violence, with no indication yet of which side will win.

History, unfortunately, does not teach that economic common-sense always outweighs the force of ideological fanaticism. There can be little doubt that the best future for the South African mining industry lies with a progressive and negotiated movement towards a multi-racial political system; little doubt, either, that the kind of physical crisis the extreme black nationalists seek would damage the industry, perhaps fatally, and so plunge all southern Africa into unimaginable poverty.

Economic facts are not always paramount; but they do speak loudly. Mining gives jobs to eight per cent of the South African employed population, a total of over 700,000 – more than in the US, Canada and Australia put together. A quarter of a million of them, all black, come from outside the Republic, from Botswana, Lesotho, Malawi, Mozambique and Swaziland, and their earnings are critical to the economy of these states. For its size, South Africa is the largest depository of minerals in the world, and exports more than eighty per cent of its production. It is the world's largest supplier of gold, platinum, gem diamonds, chrome, vanadium, manganese, andalusite, vermiculite and certain asbestos fibres; the second largest exporter of uranium and antimony, and among the top ten suppliers of coal, fluorspar, nickel, copper, tin and silver. It has eighty-six per cent of the manganese ore. The South African mining industry is the most versatile and efficient in the world and no one stands to gain by its destruction, with the sole exception of its chief competitor, Soviet Russia. Between them South Africa and Russia have sixty-eight per cent of the world's gold. Less commonly known is that they share ninety-nine per cent of the world's platinum, ninety-seven per cent of its vanadium, ninety-three per cent of its manganese, eighty-four per cent of its chrome, and high percentages of many other rare and important minerals.

Certainly, South Africa has little to fear from boycotts or embargoes. Such measures, when directed against a mature industrial power, by promoting import-substitution, tend to strengthen rather than weaken its economy, often in unsuspected ways. Thus, the United Nations embargo on supply of arms to South Africa has led to the creation of a large and increasingly sophisticated indigenous arms industry. Drawing on the unique technological resources of its mining industry, South Africa now rivals the US and Sweden in the manufacture of advanced conventional explosives and is a leading specialist in mine-resistant armour and vehicles. It exports these products all over the world, notably to black African countries; so that the UN which once enjoined the world to cease exporting arms to South Africa, was driven in 1985 to ask it to ban South African arms imports. The ban on oil exports to South Africa led to comparable results and especially to the further development of a synthetic liquid fuel industry. Other examples could be given. The 'disinvestment' campaign, launched by

Coal in transit from the Apex Greenside colliery for stockpiling at the nearby train-loading terminal.

South Africa's more extreme critics in the winter of 1984–85, is also likely in the long run to foster its economic independence by forcing it to develop its own financial markets. Certainly, so long as South African investment remains profitable, Western capital will find its way there.

The oil boycott and the development of liquid fuel from coal had an unforeseen and dramatic effect on South Africa's coal industry. Its coal reserves had always been known to be very large: that was why it was thought worthwhile to increase SASOL, the publicly financed synthetic fuel concern which turns coal into oil and petrol. But in the process the possibilities of large-scale exports of coal were explored: in particular, negotiations were opened with Japan, regarded in the trade as the most difficult but also potentially the most lucrative customer. Only fifteen years ago, South Africa was a nonentity in the international coal trade. The initial target was 2.5 million tons annually.

GFSA was able to play a part in this venture because its 1960 acquisition of Anglo-French gave it Apex Mines, including the Greenside colliery north-east of Johannesburg. Greenside has three different seams and three different products: coal for steel, and for the domestic and export market, the latter of two brands – low-cost coking coal for Japan, and steam coal for power generation and cement factories. More recently GFSA has acquired Clydesdale, with mines at New Clydesdale in the Transvaal, Coalbrook in the Orange Free State and an interest in the Matla mine. Altogether this gives GFSA just under ten per cent of total South African coal production and puts it in the major league. Hitherto, Anglo-American and Gencor were the biggest producers of South African coal. In 1984 Drury Gnodde, GFSA Deputy Chairman and Executive Director in charge of coal and base metals operations, became the first GFSA chairman of the Transvaal Coal Owners Association (TCOA), the joint marketing body.

GFSA coal properties are a mixture. The New Clydesdale operation is highly profitable. Coalbrook is a different case, as my diary of a visit makes clear:

Coalbrook is old, eighty years, undercapitalised and has only one customer, the nearby two-unit power-station complex which may disappear. A quarter-century before GFSA acquired it, on 21 January 1960, it had a terrible disaster, 435 lives lost. But now its safety record is the best in the industry. Badly needs new shaft, as men descend by old shaft, go in train five kilometres, then three kilometres by diesel, then have two-kilometre walk before working. Manager told me: 'We would have to invest R25 million to modernise it. So far as reserves are concerned we could go for another 100 years. Have 250 million tons of saleable coal. But it is low quality. Not export specification. . . . We are a cost-plus operation, so all costs are paid by Escom (state electricity authority). They must provide any capital expenditure and they won't pay for a new shaft. But to close would be a pity as the infrastructure of the mine is of a very high standard, completely mechanised, and safety top-grade. It has the best training facilities in the whole of SA coal industry.' GFSA bought the Clydesdale company because they were established operations that would enable it to broaden its technological base and expertise in the coal industry. Another factor was to enlarge its export quota, what they call 'the dowry'.

The last point is important because the potential future profits in exporting South African low-cost coal are great. The TCOA invested huge sums in the development of its coal-export capacity. It guaranteed the revenue for a new 500km railway line to the coast and at Richard's Bay built the largest and most efficient terminal in the world. It will be, when at full capacity, fed by trains running continuously, each only fifteen kilometres apart, round the clock, 365 days a year. Each train is two kilometres long, with 200 trucks, and carries 17,000 tons. They have to be loaded in under four hours. All this was created in less than ten years. By 1984 Richard's Bay was capable of handling 150,000 tons of coal a day and loading 500 bulk carriers annually. Coal exports jumped from the initial 2.5 million tons to ten million and, in 1985, to an (estimated) 34.5 million. The eventual export target is between sixty-five and a hundred million tons. Gnodde told me:

A long-wall coal-cutter underground at Clydesdale. Apex and Clydesdale were merged with effect from 1 January 1986 to form Gold Fields Coal.

> The coal resources of South Africa are enormous. We have 250 billion tons good enough for export. Transport by rail is cheap. But then the coal companies guaranteed the line and put in the ultra-modern terminal, rapid-loading bays and port. By comparison, Australia and the US are old-fashioned. Our coal is very competitive at the Richard's Bay quayside. Also first-rate is South Africa's regularity of supply and consistency of quality. We tell them: if you send your ships to pick up 100,000 tons on 1 July the coal will be there and of exactly the quality we have contracted for. We know how to cope with disasters without interrupting supplies. In February 1984 the Da Noina cyclone from the Indian Ocean broke on the Natal Coast. It not only destroyed millions of acres of agricultural land but dumped twenty-five inches of rain within twenty-four hours on the port. This meant seventeen tons of water in each coal truck at Richard's Bay. The weight of water would have broken the tip-arm of the loaders, so we marshalled men to drill extra holes to drain the water, and we didn't miss a delivery although we had 2,500 rail trucks tied up in the yards. Why, TCOA now even exports coal to America.

Pouring molten tin at the Rooiberg refinery.

GFSA's major investment in coal is only one aspect of its efforts to spread its mining base in South Africa. Some eighty per cent of the company is in gold. It built up its interests in coal and base metals both as a prudent diversification and to express the group's function to develop new mines, because gold itself cannot always be found. The mines came to it through acquisition from 1960 onwards. To balance its seven gold-mines it now has seven coal and base metal companies, having taken over the running of O'okiep Copper Company Limited from Newmont Mining in October 1984. GFSA is now South Africa's primary producer of lead and zinc and a major producer of coal and copper. But Gnodde emphasises that the object is not prestige or asset size:

We have always believed in cash flow as the most vital aspect. That means profits. We are not interested in the strategy followed by some US firms which are only concerned with growth of assets – that way, the money all goes to banks in interest charges on the money you have to borrow. . . . It will be a long time before this division makes the profits of Driefontein, but then we didn't have to make a comparably huge capital investment either. In fact our investment was R40 million. The figures are not big but provide reasonable profits.

Mining base metals for profit is not easy at a time of low world prices. When GFSA took over the running of the O'okiep Copper Company from Newmont, the mine was producing 20,000 tons of copper a year and losing R15 million. It has its own primary smelter but sells blister copper which goes to another refinery. The aim of the transfer to GFSA was and is to reduce losses, made easier by the decline of the Rand against the US dollar. GFSA produces tin at a modest profit, from its small Union Tin mine (300–400 tons a year) and the much bigger one at Rooiberg. Together they produce about 2,000 tons of tin a year, 1.5 per cent of world production and constitute essentially South Africa's main tin provider. In 1978–79 it put in its own smelter for high-grade concentrates, and now markets refined tin in Japan, Continental Europe, Britain and eventually, it is hoped, in the United States. GFSA also administers Zincor which has its own electrolytic zinc smelter, the only one in the Republic which refines zinc concentrates from its own mines and other sources, and produces 100,000 tons of zinc a year, enough to supply the South African market. Most of it is used in galvanising corrugated steel roof materials.

Much the most interesting GFSA operation in base metals is the Black Mountain mine in northern Cape Province, on the edge of the Kalahari Desert. This is among the most complex and technically advanced mines in the world. It illustrates one of the most fascinating aspects of South African mining history: you start looking for one metal and end by finding something quite different. Thus, GFSA's Apex started looking for gold and found coal. Rooiberg looked for copper and found tin. In 1970 Phelps Dodge Corporation drilled for diamonds on the Aggeneys Farm north-east of Springbok and in the next three years intersected three orebodies (Black Mountain, Broken Hill and Big Syncline) containing lead, copper, zinc and silver. Phelps Dodge spent R15 million for a fifty-one per cent interest. It also provided an interest-deferred loan of R35 million and a further R111 million was raised by loan finance.

Dru Gnodde told me:

We planned the mine and brought it into production. It is very successful, but its capital structure is not quite our style since it must have been the most heavily geared company we have. Altogether it cost R180 million to bring it into production and the company is not yet earning profits for shareholders since R25 million a year goes on interest charges.

GFSA now owns fifty-five per cent and has made fierce efforts to reduce the burden of debt. Moreover, the mine was planned at the height of the silver

OPPOSITE GFSA took over responsibility for the O'okiep Copper Company from Newmont Mining in October 1984. GFSA is now South Africa's primary producer of lead and zinc and a major producer of coal and copper.

Black Mountain in northern Cape Province on the edge of the Kalahari Desert is one of the most technically advanced mines in the world. It produces lead, copper, zinc, and silver.

boom, and since then base metals as well as silver have fallen in real terms. 'Things were quite grim two or three years ago,' I was told. But the mine has survived and prospered thanks to relentless cost-cutting, providing another example of my adage that the best test of entrepreunerial spirit is breasting a wave of recession. (The capital has been restructured and a maiden dividend was paid in 1986.)

My diary records:

Up 4.45. Drive to small airport of Lanseria north of Johannesburg for 6.30 flight to Black Mountain. Three hours, route WSW. Over Vryburg, seat of Soldiers' Republic in 1880s, then Kuruman, old missionary centre 1820–40, where the Rev. Robert Moffatt had his HQ. This has an 'eye' where there is a rush of water to the surface, which has produced superb water-gardens. In 1841 David Livingstone arrived and in 1844 proposed to Moffat's daughter Mary under an almond tree which is still there. He made it his HQ for his northern journeys. Then over Sishen, site of huge open-cast iron mine, Upington, once bandit centre, and 600-foot high Augrabies Falls, fifth highest in the world. As we neared Aggeneys, the land became orange-brown, with ridges of fierce hills, many with zinc and iron deposits. There is no water apart from the Orange River, with cultivated banks, like the Nile running through the desert. Saw the mine's water-pipeline from the Orange River, running through gap in the hills. Exciting and risky flying into airstrip in bowl of grim mountains. Aircraft taxied exactly into sun shelter, with only an inch or two to spare on each side.

Visited mine and plant with Peter Kinver, Assistant Manager, Tim Twidle, Plant Superintendent, and Hennie van Aswegen, Resident Engineer. Mine has incline for main access. Drove to levels 6–7, then 10–11. Elaborate system of rubber venetian blinds, opened by drivers pulling hanging ropes, which seal off sections and conserve cool air. Watched cut-and-fill mining. . . . Then to tops of silos, where ore is taken by trucks to be dropped down to crushing level – the primary crushing is done at bottom of mine. Silos are metal-lined circular holes and huge steel plugs come down to prevent dust rising to our level. Essentially, you mine upwards, but everything goes down to bottom of mine, where it is broken between steel jaws of a crusher, then put into 10-ton skips and lifted out to surface up the main shaft. It is taken by conveyor belt to secondary and tertiary crushers. Then into mill, where it is treated in vast steel drums.

I then watched concentration process. This bubbles everything up and, when it settles, it is at this point that the varying properties of metals separate one from another. The division here is threefold – copper, lead, zinc – with silver in both copper and lead, despite the fact that it is, after lead (at present), the second most valuable product of the mine, followed by zinc and copper. The control-room of the flotation plant is remarkable. A Courier 330 Analyser automatically takes samples of all flows every six minutes, X-rays them and reports quality; Proscon 103 and Proscon 105 computers control the main console. In the middle sat one girl, alone. It is run by Control Supervisor and three ladies working in eight-hour shifts. A visiting mining consultant, who was with me, said it was the most advanced control room he had seen anywhere.

After separation, I saw the various concentrates de-watered and dried out, then dropped into three enormous storage sheds. This year the mine produced nearly 95,000 tons of lead concentrates, 8,000 tons of copper, 25,000 tons of zinc and 106 tons of silver contained in the lead and copper concentrate. From the sheds, sixty-ton trucks take the three concentrates to Loop Ten for train-transport to the port. All copper and lead go overseas, and the zinc to GFSA's zinc smelter. This mine is on the frontiers of advanced technology, but is run in a practical, no-nonsense fashion with the stress always on cost-savings. For instance, by substituting a local sand-mixture for imported concrete they have made big savings because of the colossal quantities used. What is impressive is the way management solve problems empirically as they come up, often inventing quite new solutions, partly because they have a complex metallurgical operation raising novel difficulties, partly because they are so isolated and have to be self-reliant, away from bosses – despite daily plane link – and partly because they are determined to make a profit, despite the sorry state of the metal markets, by cutting costs. All mines are individuals, almost like people. Black Mountain is a most unusual specimen. Kinver showed me different blasting techniques, very complex with many variables, which can make all the difference between profit and loss. He says: 'Engineering theory can give you the foundations but you have to know your own rock and live with it.'

Casting zinc ingots at Zincor.

Ten years ago, the Black Mountain was nothing: now it has a turnover exceeding R100 million a year. Black Mountain is an isolated community and as often happens, isolation breeds enthusiasm. It is GFSA's policy to lock the gold and base metals divisions together by shunting people between

A jumbo drill at the face in the Black Mountain mine.

the two every three to five years: thus a man may serve first on Driefontein, then at Black Mountain, then at Rooiberg – or the other way round. This develops the group spirit and identity.

Gold Fields of South Africa's future cannot be separated from the future of the country as a whole. Both will eventually have to come to terms with the black majority population, as workers and as citizens. Bernard R. van Rooyen, GFSA director in charge of Corporate Finance, stresses:

> The labour dependence of the South Africa mining industry is enormous, especially in gold because of the narrow seams. It employs 500,000 black workers. The mining industry here is technically the most efficient in the world. But it is often not economic to use machinery rather than hand-controlled, fairly elementary technology. Admission of blacks to unions is only five years old. Fortunately there is a long history of militant white unionism, so there is a high degree of legislation governing the resolution of industrial disputes and a series of legally-binding steps you must go through before you can strike. Unions are legally prohibited from calling political strikes.
>
> In economic terms, the mining industry has a good future. We have capital, technical expertise, excellent labour relations and plentiful reserves. The wave of economic nationalism is on the wane – exchange controls, banking controls and so on, are being eased. Some outfits may go to the wall, because hitherto the government bailed you out. Our position is very strong. We can last thirty years if we never opened another mine. But this is no head office hothouse. We grow up in an

operational atmosphere. The South African mining house system is unique in the world – it explains why the gold rush didn't die like California's or West Australia's. It provides for very great delegation of operating authority. But there are centralised engineering services and so on. So we keep together good teams of experts. GFSA protects operating companies from aggressive takeovers.

GFSA's new headquarters in Fox Street exude confidence, based primarily on the company's immense financial strength and its virtual freedom from any debt. It is looking hard for further expansion both geographically and in terms of mining products. Surveying the future with me, Robin Plumbridge laid down the guiding principle:

Our definition is mining, whereas London's is natural resources. We are in mining and beneficiation. I doubt if we would want to get into the construction side. London agrees with our aims. In 1960 we were under enormous pressure from Head Office to get involved in manufacturing. Now, more and more, there is a common purpose in the totality. Oil? Not interested in South Africa – the chances are low, the risks high, only minor quantities have been found. Energy? Yes – we are in coal. Another possibility is the chemical field, though we have taken a big share in SASOL itself, as it's technically superb in this area.

If you're in mining, the future lies in finding new mineral resources. That's what we mean by new business and acquisitions. The great quantum leap comes through your own home-grown projects. Acquisition is a sideline – necessary for strategic purposes – but the real stuff comes from new projects. The phenomenal decision taken in the 1930s to open up the West Wits Line gave us a marvellous bottom-drawer situation, and we are still drilling in that area today. But in one respect it was a disadvantage. Having made the great discovery, the group went into its shell and didn't go for new areas and ideas. Not enough was ploughed back into exploration. In the last few years we have recognised that our bottom drawer – our larder if you like – was getting pretty empty. So now we are pursuing an aggressive policy of exploration and acquiring mineral rights – gold, coal, platinum and so on. This is why the great financial strength acquired in the last five or six years is so important. It has enabled us to set aside large sums for exploration, buying rights and long-term projects. It is like R & D in manufacturing. We have been able to make strategic acquisitions out of retained profits without worrying about our basic earnings structure.

This is how Chairman Plumbridge sums up the future: 'The benefits will not come in the short term. It will take fifteen years. The real benefits will come in the late 1990s.'

Drilling 6-inch holes prior to blasting an ore pass at Black Mountain.

A miner removing waste from a conveyor which feeds the primary crusher underground at Black Mountain.

3 Gold Fields' Operations in America

GOLD FIELDS has a knack of running its affairs from unusual, historic or even spectacular headquarters' buildings. The New York Offices of Gold Fields American Corporation, on the top floor of the Helmsley Building, 230 Park Avenue, is perhaps the most satisfying. It was designed in 1929 by Warren & Wetmore for the Vanderbilts as the head office of the New York Central Railroad. Its tracks once ran in the open up Park Avenue, before the creation of Grand Central Station pushed them underground. Number 230, though built only a year or two before the Empire State, seems to date from the opulent, robber-baron era before the Great War. Today it is overshadowed by the vast Roth-Bellushi-Gropius Pan-American building behind it; but once it served as an apse of grand proportions to culminate Park Avenue, its main cornice line at the 13th storey binding it to the tops of the flanking buildings on either side of the Avenue, and its soaring tower and exotic cupola visible for miles around. Its lobby blazes with scarlet and ormolu. Its gilded lifts, with their ceilings of painted feathery clouds, might have been made for Versailles. Its top, boardroom floor ought by rights to be termed its *piano nobile:* now superbly restored, it recalls Vanderbilt glory at its apogee and actually served as the vainglorious setting for the supreme council of the Mafia in the film *The Godfather.*

Perhaps it is a boardroom to encourage *folie de grandeur*; certainly it witnessed some dramatic deliberations in the early 1980s. As we have seen, Gold Fields' involvement with the United States goes back to the time of John Hays Hammond in the early years of the century. His advice, that a company like Gold Fields must acquire substantial assets in America, was valid then and remains valid now. America is the largest and richest economy in the world; in some ways the most open; it is the core of the capitalist system; not least, it abounds in natural resources and minerals of all kinds. It would be absurd for an international mining finance house like Gold Fields not to operate in this uniquely favourable environment.

Yet the exhilarating business climate of the United States is also a highly dangerous one. It oscillates at great speed between optimism and pessimism, and is subject to unpredictable and devastating legislative gusts from Washington. It was such a gust of new environmental laws from Congress which wrecked Gold Fields' first really big United States venture, American Zinc, leaving it with a $30 million write-off and a tempting tax-credit. The subsequent large-scale venture in the American manufacturing sector, with

OPPOSITE GFMC has its head office in the Helmsley Building, 230 Park Avenue, New York which was designed in 1929 by Warren and Wetmore. The Newmont head office is in the Panam building behind.

Since the interviews reported in this chapter the whole of the North American industrial group has been sold to a management group. However, the lessons from this episode have been so marked that I thought it important to retain it rather than omit the history of this period.

its uneven and in one case disastrous results, has already been briefly described. We must now look at one or two aspects of it in more detail, before examining Gold Fields American Corporation operations as they are now.

The early 1970s were dominated by the need to create profits to set off against the tax-losses arising from the sale of American Zinc. This involved buying companies and some of the purchases were shrewd. The first was George E.Failing, known as the Cadillac of the mobile drilling equipment industry, specialising in water-drills and vibrator trucks; it was bought from American Standard – which almost immediately regretted the sale – and performed very well throughout the 1970s. Next came Brown-Strauss, steel distributors and stockists. This again was brilliant timing. In the first two years of Gold Fields' ownership it made profits of $12 million, almost as much as the purchase price. Another good buy was the Steel Service Company and Knoxville Iron in Tennessee, which produced healthy cash-flows. A fourth winner was Unimet Corp, a mini-conglomerate in steel distribution and special shapes and pump-parts, very profitable in the 1970s, and still profitable today, though less so.

With hindsight and seen from the perspective of the mid-1980s, it is not hard to pick holes in the strategy. Gold Fields was buying into steel and into the oil exploration servicing industry, areas in which it had little or no experience. It was, moreover, buying (with excellent timing) into a boom in both cases. As Bob Krones, then Vice-President in charge of legal affairs for Gold Fields American Corporation, put it to me:

> A company which tries to do everything is making a mistake and it is at the top management level that the lack of real understanding is felt. The idea that, even if top management doesn't understand the business, you can buy yourself junior management who understand it, is quite wrong. A company may do well for a couple of years. But a crisis is sure to come sooner or later and it is then that the top management needs to be good and understand the business thoroughly. It is a bad idea for firms to engage in activities that their senior managements do not know about or feel comfortable in – and where they may react badly in a stress situation. In fact, if things go badly for a bit they tend to over-react.

That indeed was the lesson of American Zinc.

Joe Davis, GFAC Vice-President for finance, makes the further point that buying into steel distributors with large stocks on the eve of a steel boom was a form of speculation in commodities:

> This was not appreciated. Of course in this case the speculation was initially successful but there are natural downs as well as ups. I'm not sure they did have a clear plan to sell as soon as the tax-loss was used up. In any case they soon forgot it. It is very hard to bring yourself to sell a company making good profits. They found themselves empire-building with the additional rationale – then – of getting out of South Africa into the safe Western hemisphere.

In business there is often a very narrow dividing line between great and solid

A barge-mounted Skytop drilling rig capable of prospecting for oil to a depth of 30,000 feet.

success and spectacular failure – which in retrospect looks reckless. In an enterprise like Gold Fields which positively encourages risk-taking and is unwilling, in the words of Sir Winston Churchill's famous minute, 'to penalise mistakes made *towards* the enemy', a daring senior executive with an excellent record poses a real dilemma. As Peter Roe, a former Company Secretary of CGF, remarked to me: 'One of the greatest problems in this company is to know when to control a successful risk-taker who is going over the top.'

Between 1979 and almost the end of 1982, however, faults in corporate structures led to serious problems. By the late 1970s, the expanding area was oil exploration. According to Vincent Filippone, 'In 1980 we were looking for a manufacturing facility in oil. Skytop Brewster was owned by Texas International, which needed cash at the time. It makes big, fixed drilling-rigs, skid-mounted as they are called. Skytop was looked on as a "turnabout situation". Gold Fields felt, though it had never been in the business, that it had the skills to turn it around.' Joe Davis put it a little differently: 'In April 1980 energy was the top thing. Skytop Brewster was for sale at $60 million and it seemed a cheap way to get into the energy business. It *was* cheap, but its management was the problem.' According to Filippone, Skytop never had the management skills to achieve a permanent turnaround. 'They were inexperienced operators. They came along with the purchase. It's usual to give a five-year contract in these cases. I suspect Skytop management had not been there very long when Gold Fields acquired it. Their inventory systems were bad.'

Skytop-Brewster bought many of the parts for its rigs from other manufacturers. In turn it sold its completed rigs to a multiplicity of oil exploration companies, most of them small. The oil industry, and therefore the businesses which supply it with equipment, are highly cyclical. There is an accelerating build-up to a climax, then a sudden, devastating downturn, usually without much warning. As Filippone puts it, 'When the bottom comes in the oil market, the phone doesn't ring.' In this business, firms ordering rigs do not make deposits. Cancellations are frequent. A full order-book does not necessarily mean much. The art of the rig-supplier is to have a low inventory and little debt when the downturn comes.

By acquiring Skytop in April 1980, Gold Fields seemed to have bought cheaply into the beginning of a boom. The rig-count, that is the number of oil exploration rigs in active use, went rapidly from 2,000 up to 4,000 and then headed for the 5,000 mark. It was a bit like the stock market. A rig could be sold instantly it was assembled, at top prices. So Skytop expanded capacity and ordered parts in huge quantities.

From June 1980 up to December 1981 it was marvellous, [Joe Davis told me]. You were making so much money, so quickly, it almost didn't make sense. Some people in London were suspicious. They wanted to sell it now. Others wanted it to become a portfolio company, issue paper and so on. Our investment in Skytop went from $60 million to $150 million and then up to $200 million in eighteen months. What was happening was hidden to view – long lead times, and you were at the vendors' mercy. On the one hand, we were buying from the vendors all this material. On the other, we had a full order book. Suddenly, the rig count dropped. Everyone cancelled, leaving Skytop with enormous capacity, a huge inventory and no customers whatever. We tried to get back to our vendors. They said – no, you can't cancel on me.

With hindsight, it seems a clear case of *hubris* followed by *nemesis*. Joe Davis says: 'It makes no sense in retrospect and it made no real sense at the time.' Art Johndrow, the very experienced President of Gold Fields' other company in the rig business, George E. Failing, told me he took a less critical view. Here was a case where the actual experience of living through a fierce cycle was important. 'We learnt a lesson from the 1957 collapse of the rig market – you must have a big spread of products, and a big spread of industries you supply. In the 1982 crisis, things moved with terrifying speed. Skytop's three main competitors were all in the same trouble. We were fortunate. We quit buying. We quit speculation. We cancelled and returned material. I rang up the suppliers myself and cancelled. But a month later it would have been too late.'

Skytop's position, when the phone suddenly stopped ringing, was vulnerable because it was the wholly owned subsidiary of a famous and wealthy international mining finance house. It could not take advantage of legal remedies to evade its obligations to parts-suppliers. As Davis puts it:

> Sure, we could have gone into Chapter Eleven (voluntary liquidation) and read in the *Wall Street Journal* the next day: 'Subsidiary of Gold Fields Goes Bankrupt'. Trouble was, we were selling to small people – there was no sense in suing them anyway. We were caught in a classic squeeze. In one month alone $35 million went out. We ended up with a $150 million inventory. In December 1982 Gold Fields created a $151 million reserve (£87 million at the then exchange rate) to cover all losses. God, did they get it over, rather than fork out £10 million a year for a decade. It was Humphrey Wood who cleared up the mess.

I have described this episode not so much because of its intrinsic importance, for all international firms with a wide spread like Gold Fields meet setbacks on this scale from time to time, but because it pointed to a weakness and confusion in overall strategy. There was a lesson to be learnt; and it was learnt. The effect on Gold Fields was to send it back to its origins: to mining and to the new broader concept, developed in London by Rudolph Agnew, of Gold Fields as a developer and beneficiator of natural resources.

One consequence, in particular, was to upgrade the importance of gold-mining in Gold Fields American plans. This was good news to William K.Brown, President and Chief Executive Officer of Gold Fields Mining Corporation, which controls the company's wholly owned mining and exploration activities over the whole of the western hemisphere. Brown has

Bill Brown, President and Chief Executive Officer of Gold Fields Mining Corporation.

spent his life in mining. He was with Phelps Dodge and developed the Black Mountain project; it was he who made the arrangements with Bernard Van Rooyen to bring in GFSA. He had agitated for Gold Fields to spend more money and effort on doing what it knows best – finding and recovering gold. In 1981 he told a gold seminar: 'The US has to be regarded as a favourable geological environment for gold, particularly the western states'. Newmont's discovery of the Carlin deposit in Nevada in 1963, he added, 'marked the start of a new era in US gold exploration'. In the twenty years up to 1981, at least $300 million and probably more was spent on gold exploration in the United States. The exploration cost averaged about $13 an ounce. During this period an estimated 700 to 750 tons of recoverable gold were found, with Newmont's Carlin (130 tons) topping the list, followed by Homestake's Napa Valley mine. Most discoveries, like Gold Fields' own Ortiz mine at Cerrillos in New Mexico, are small, more in the region of ten tons. Most of this is sub-microscopic gold, which the techniques of the old prospectors did not allow them to recover, though they usually found the areas where it existed.

Brown regards the rise of the gold prices in the 1970s as the modern equivalent of an old-fashioned gold rush, provoking intensive exploration for gold all over the world. The best areas to examine are the Precambrian Conglomerates, of which the Witwatersrand Basin is the classic example. But these conglomerates are also known to occur in Ghana and Brazil – where they are actually being mined – and in India, Australia, Canada and the United States. Brown thinks the practical chances of another Witwatersrand being found are 'very low', particularly given the 'political restraints to exploration in certain areas'. But substantial discoveries are pretty certain provided the exploration effort is adequate. He reiterates: 'For a mining company, exploration is the equivalent of R&D in manufacturing.' He told me that in his view Gold Fields in the past did not do enough exploration for gold in America. But now 'I think Gold Fields has got its philosophy right – one reason I switched to come here.' It has become a large spender on exploration in the Western hemisphere, about $20 million a year.

Brown took me through the prospects for the Americas. Of course, those outside the US were influenced by the political climate:

> In Canada, since they've booted Trudeau out there is a less anti-foreign business atmosphere. In Chile, the government is very pro-foreign investment, so we are there. We have interesting prospects in Chile – we might have a gold and silver mine. In Peru, it is very attractive geologically but in the last eighteen months the political climate has deteriorated to the point where we have wiped Peru off the list. The Argentine too.

> As for exploring in America, environmental restraints are now a major factor – vital. Parts of California and Minnesota, for instance, are useless, whatever we were to find. There are parts of California we wouldn't touch because of environmental laws – legislation can be put on ballot which effectively bans all mining. In the area of water and air pollution, things have become steadily more difficult. Cyanide, used in the gold

leaching process, is an especially emotive element. It's illogical really, since the very weak solution we use is biodegradable in the climate. But it has a bad name. California is the most difficult. Nevada is much easier. But everywhere, we move with a phalanx of lawyers all the time.

In some respects, however, the worst of the environmental pressure is over. There is more realism. We have a younger group of mine managers not hindered by old-fashioned attitudes. They take environmental factors in their stride, as a fact of life – no point in getting mad at it.

Of Gold Fields' two American gold mines, Ortiz, though much the smaller and now almost worked out, was very popular with mining engineers. Robin Hickson, who opened it up and was mine manager from 1978–83, says: 'Ortiz is one of the most desirable spots to live in if you're a mining engineer.' It has a desert climate and, like Johannesburg, is about 7,000 feet in elevation, a plateau with weird mountains sticking up around it. It is between Sante Fé, with its live theatre, opera and artistic community, and delightful Albuquerque, with skiing nearby. It has always been popular with gold miners, and it is claimed that local miners who live there are the direct descendants of the original Conquistadors. The problem has not been finding gold but recovering it. The discovery of a nugget there in 1828 led to the first gold rush west of the Mississippi: 4,000 came to the Ortiz area. It got its name from José Francisco Ortiz, to whom the Spanish

A blast at Ortiz which was the first mine developed by GFMC. It was relatively small, but highly successful, and produced a total of 8 tonnes of gold between 1981 and late 1986 when it closed because ore reserves were exhausted.

Gold in solution flowing from the leaching pad to the treatment plant at Ortiz.

government gave a mine grant of ninety-five square miles. But the 1848–49 California rush took all the miners away. There was another bout of mining in 1864–80 and again in 1899–1903. Thomas Edison, who stated in 1899 that Ortiz contained 'gold worth $800 million', had a go with his static electricity separator method; but this proved a classic case of a process working well in the laboratory but not on the ground. Gold Fields began seriously prospecting for gold in America in 1971 and took a lease on Ortiz in 1973. After five years of studies, with the gold price rising, full-scale development of the mine began in 1978, with production starting in February 1980. By 1984, over 900,000 tons of ore were being leached per annum, with a production of around 40,000 ounces. This put it among the top ten gold mines in the US.

Hickson told me that Gold Fields was allowed to develop Ortiz because it

Sprinklers in operation on the leaching pad at Ortiz. The weak cyanide solution dissolves 80 per cent of the gold content of the ore over a ten-week period.

BELOW Scraping gold foil from stainless steel plates after electro-deposition at Ortiz.

met environmental standards in three respects. Firstly, the actual design of the mine incorporates anti-pollution measures: it has a liquid closed circuit, with no water disposal. Secondly, there is constant ground-water monitoring. And thirdly, on the positive side, its re-seeding plans and the engineering of its drainage system are specifically designed to benefit wildlife and indigenous plants.

Ortiz is an open-pit, the tiny particles of gold disseminated in a hard siliceous volcanic rock are recovered by chemical leaching. The method is worth describing in a little detail because although, as Bill Brown says, 'it is basically the simplest technology there is' and each component has been long known, it is also true, as Hickson told me, that they were not put together until the 1960s, 'so it is essentially a new process'. At Ortiz thirty-foot holes are drilled in horizontal rock benches twenty-five feet high and then blasted; each ton of ore contains less than one-twentieth of an ounce of gold. The ore is picked up by ten to twelve ton capacity front loaders, dumped into a fleet of fifty or eighty-five ton trucks which feed a primary crusher producing per day 3,500 tons of rocks under six inches in diameter. The fine crusher then reduces them to the optimum size of 5/8ths of an inch (both crusher plants have dust suppressor systems). A conveyor takes the crushed ore to an asphalt leach pad and stacks it in a 23-foot-high pile, which is sprayed with weak cyanide; the solution flows by gravity to a ditch and sump. After several weeks of leaching the now barren rock is picked up again and dumped on the waste heaps. The solution or 'leachate' passes through five carbon adsorption towers, eight feet in diameter and six feet high, at a rate of 750 gallons a minute, the five-tower passage taking ten minutes. Caustic is used to strip the carbon particles of gold. The gold-bearing caustic is then piped into a two-stage electrolytic cell process, which plates the gold onto steel wool and then onto stainless steel sheets. The gold foil, known as Doré and containing some silver, is then scraped off by hand and conveyed by armoured car to the refinery for purification.

Mesquite in the California desert is a much bigger proposition, with three or more times the potential output. Gold Fields spent $10 million in 1981–84, completing its feasibility study in March 1984, with the decision to proceed with full-scale development following a couple of months later. Again, it is old goldmining country, now being brought back into production. Gold Fields' deposit is known as the Big Chief. Nearby are Newmont's American Girl and Chemgold's Picacho, the latter a leach operation on low-grade ore left behind by the old timers. As this is California, the environmental obstacles were much higher than at Ortiz. Indeed in this area many thousands of acres are at present unavailable for exploration either because they are military ranges or because they are designated as 'Wilderness Study Areas' or 'Areas of Critical Environmental Concern'. There is also the problem of getting the land rights under California's complex and ambiguous 1872 Mining Law. Bill Brown told me:

Our strength in this particular case was to be able to put together the land position. We have an in-house land and legal department. As a

consequence, we can develop and control our strategies well. Other mining firms depend upon big law companies. So we were able to put together a land package. Then there is the permitting process. We have been doing it for a year. It's fortunate that the deposits are in a part of California which is pro-business.

Indeed, the votes of locals are decisive, and from the start the Gold Fields' people involved in the project have done everything in their power to get

Mesquite in California was brought to production by GFMC early in 1986. It is now producing some 5 tonnes of gold annually and reserves are sufficient for more than 15 years.

themselves accepted. The largest major permit came through on 30 January 1985, just before I flew to the California desert. My diary continues:

> Took Sun Air flight to Imperial County. Rich valleys, then mountains with Mount Palomar telescope on right, then desert under crystal sky, with huge salt lake, Salton Sea, 235 feet below sea-level. Landed El Centro, fifty-four feet below sea-level, eleven miles from Mexican border. Hickson drove me to the mine. He is from Lincolnshire, of farming stock. Here they grow broccoli, cabbage, cotton, lettuce, beet, onions, alfalfa, six to eight crops a year of some things. Southern Pacific has line only six miles from Mesquite. Drove through lunar landscape of sand-dunes, much relished by dune-buggy fans, a typically California craze. At the Glamis Beach Store signs proclaim 'The Only Beer for 19 miles' and 'The Sand-Toy Capital of the World'. Hickson buys his petrol here, despite the prices, and chats up the people as part of his public-relations campaign. He thinks that 'once things get going it will probably make more money than the mine'.
>
> The mine is in the Chocolate Mountains, between Chocolate Drop on the left and Brownie Hill on the right. They really do look like mounds of chocolates. Hickson has only eighteen people working at the mine at present – the big rush is about to start. He hopes to begin production in January 1986. Big Chief will be the initial pit, one mile by half a mile, then move over to the Lena Pit, process thirty-six million tons, then move east. They will end up moving ninety tons of rock for every ounce of gold. Showed me primary and secondary crushers, and agglomeration plant. Then onto the heap leach piles, layers fifteen feet high, sprayed by garden-type hoses with solution of 0.1 per cent sodium cyanide, the rest water. The carbon they use in the recovery process is burnt coconut shells. The difference between here and Ortiz is that Mesquite is big enough to make its own gold bricks from the Doré foil. [Commercial production began in March 1986 and the results have confirmed the feasibility study.]
>
> Essentially I was seeing the pilot plant. We went over the whole area of operations. As Bill Brown says it is a simple process but has to be worked with great attention to detail and continuous monitoring of ore quality. Costs can be lowered very substantially by expert attention to minor factors, based on experience. And costs are all-important in this low-grade ore operation. Hickson says 'We can mine profitably at the present gold price, though it's not pleasurable, or even below – but in that case we'd have to do things differently, change the stripping ratios and so on'. The technology of the mine was designed by the leading firm in the US, from San Francisco. Otherwise Hickson and his team have done it all, though as it is the apple of Bill Brown's eye he comes here often. Hickson told me: 'In fixing the layout of the mine you have to consult archaeologists, the wildlife people and the highways department, to get your "visual ratios" right. There's a ten-inch long desert tortoise we're particularly keen to preserve. This place breeds rattlesnakes, lizards, kangaroo-rats and jack-rabbits, though they're all scarce – not enough vegetation in the desert. But I've seen four coyotes this week.'

Hickson thinks the Mesquite mine has a big future. In the Chocolate Mountains they have a rig drilling for more ore: 'It looks good.' Gold Fields' land department have put together over 130 owners to make a viable development area. 'We are in a strong position here', he says; 'We are finding extensive mineralisation to the east of us.' He thinks there will be another Ortiz-size deposit there. Personally, as he puts it, 'I feel I am a very lucky fellow to have been in right at the beginning of two gold mines. I think I am lucky to have seen this embryonic mine on the very eve of its explosion into an active producing mine.'

Mesquite is one of four major new American gold mines coming on stream, including Homestakes's McLaughlin Mine in California, expected to

KNOXVILLE IRON CO'S ROLLING. MILL.

SHIPPING MARBLE TO NEW YORK.

Knoxville in 1889. Gold Fields purchased American Zinc in 1963 for $18 million which included these stone operations, subsequently sold to Asarco in 1971. Knoxville iron was acquired in 1974 and sold to Blue Tee Corp. in 1986.

Smelting scrap to produce steel
at the Knoxville mini mill.

become the largest in the United States. They join four major mines already
established and, spread through California, Nevada, South Dakota and
Utah, mark a renaissance of the American gold-mining industry. Gold
Fields' exploration division at Lakewood in Colorado, by general agreement first-class, is certainly not starved of funds, and is well placed to give
Gold Fields Mining Corporation a substantial share in this new industry.
What Bill Brown would like to see is three Mesquites: that indeed, would
make the mining division a major contributor to Gold Fields' profits. (In
November 1985 the discovery was announced of 'significant mineralisation'
at Chimney Creek, Nevada. The feasibility study on this attractive prospect
is currently underway.)

The expansion of the mining effort, however, is only one aspect of Gold
Fields' new strategy in America. Equally important is the transfer of the
investment emphasis and funds from manufacturing to natural resources.
As Richard Secrist, then President and Chief Executive Officer of GFAI, put
it, 'it is a process of divesting and investing at the same time'. While Gold
Fields is moving out of manufacturing, it is moving further into mining, as
we have seen, and reinforcing its position in construction materials by
building up the strength and reserves of ARC America, the US offshoot of
its highly successful British company.

Gold Fields announced publicly its plan to sell off its manufacturing side. That was not an easy move, for it included some first-rate businesses which prospered happily under GFAC ownership. I travelled down to Knoxville, Tennessee, to see one of them, the interlocked Knoxville Iron and Steel Service companies, which Gold Fields acquired in 1974 and then broke apart to operate as separate profit centres. Knoxville is one of those gritty, highly flavoured cities that make America such an interesting country. The steel operation goes back to 1868 and is in the heart of what is now a residential area, as though Mao Tse-tung's dream of a steel mill in everyone's back garden had actually come true. In fact it includes a magnificent garden created by the Knoxville Iron Company's one-time boss, Ivan Racheff, an expert metallurgist who was the first to develop a durable spring for cars in the 1920s; it is assiduously kept up by his successors. But when I went it was deep under snow, and in the steel mill the colours and chiaroscuro were unforgettable – blasts of flame shooting into the grey air, red-hot strips of steel writhing and whipping about, dim weights moving with menace overhead and the sparks shooting and hissing into the snow and ice.

Jim Pitts, the President of KIC, an old steelman – both he and his father worked for Bethlehem Steel in its great Birmingham, Alabama plant – told me he found a turn-of-the-century mill here when he took over in 1969. So he began a modernisation programme, with a continuous caster and new rolling mini-mill:

We make reinforcing bars, fifty per cent of our production. Railroad spikes. Underground support systems. So we service the construction, transportation and mining industries, quite a wide spread. Our mini-mill produces over 200,000 tons a year. We don't have blast furnaces. We use scrap. We have a machine which grinds autos into chunks the size of a coffee-cup. It separates the glass and so on. People in East Tennessee and Kentucky, being poor, buy a lot of second-hand cars, so we have a marvellous source of scrap.

But Knoxville is not a steel town. I've had to import skilled men. My workforce comes from the mountains and hills. Close to the old English in their language and slang. Suspicious of outsiders. In 1971 I negotiated a three-year contract with the local union. Fine. We doubled capacity and got a $6 million debt. Then the Federal Environmental Planning Agency began to impose fierce controls on pollution. This hit the construction industry, to which we were tied. In steel, it's feast or famine. When the market soured, we looked for a strong financial partner and found Gold Fields. We have always had a very good relationship with Gold Fields. They have always kept their promises to the banks.

In 1976 I had union troublemakers here, and we had a rough time. I had to take out injunctions against mass picketing. There was violence. I recruited new men, broke the strike and got rid of the troublemakers. This effectively saved the company when bad times came in 1982–83. The steel industry is now very competitive. You can't pass on increased costs any more. So in the spring of 1984 we decided to consolidate our two

Continuous casting of reinforcing bars for the construction industry at Knoxville.

rolling mills into one modern mill with high technology. It's now under construction and we hope to bring it on stream without losing any time. Even in this construction period we will continue to make a reasonable return. Our programme will be completed in 1986 and profits will then be satisfactory.

KIC sells its product to its twin neighbour, Steel Service, in the form of bundles of sixty-foot-long steel bars. Bill Morrow, aide to the President of SS, explained to me:

Much of our business is in Epoxy bars. Epoxy is the coating you use to stop corrosion. It comes in the form of a powder. What happens is this. You heat the bar. Then you apply a fusing-bond process. The powder penetrates the bar while it is hot. So it stops the oxydisation. We grew up with Eisenhower's interstate highway programme. About fifty per cent of our steel still goes to contractors building highway bridges. We do the coating, cut them to the customers' length and put in right-angle bits. Our vehicles haul it direct to the job-site. We ship when contractors say they're ready – we have to be flexible. The Epoxy coating fights de-icing salts used on roads and bridges to remove snow and ice. This slips into cracks in the concrete and corrodes the bars if they're not coated. Epoxy is specified by the Transport Department – and that's good for SS. Epoxy coated steel rods are also used in parking decks, because salt drops from the wheels of cars. It's needed in chemical and sewage treatment facilities which use strong chemicals causing oxydisation. Requirements for Epoxy are now moving south, below the frost-belt – even in Georgia. In Florida, Epoxy bars are now specified quite often because of the sea-water and salty air. So trade is expanding.

So it may be; and it could be argued that salt-resistant reinforcing bars are an element in the construction-materials industry, in which Gold Fields is strongly and logically entrenched, both in America and Britain. But Steel Service is inextricably linked to Knoxville Iron; and running steel mini-mills cannot plausibly be said to be within the Gold Fields' area of special skills. Equally, you might argue that mobile drilling rigs and vibrator trucks used in geophysical surveys – the main products of the George E. Failing Company of Enid, Oklahoma – are vaguely related to the work of a great international mining house. But sensible mining companies do not normally make them, and that is the real argument against Gold Fields owning Failing.

It may be a conclusive argument too; but it is the only one, for there is no doubt Failing is a superb company. I flew to its headquarters at Enid from Oklahoma City by light plane, and was met by Art Johndrow, its President and moving spirit. He told me:

George Failing started this company in 1931 making rigs. He was a strong personality. He came from the oilfields and was as tough as a boot. But in 1953 he had to sell it to Westinghouse Air Brake Company. He found he didn't like its corporate atmosphere, so he resigned. Gold Fields bought it in 1972. Under the new ownership, things have gone well. Gold Fields have provided resources for development, but we have kept the company the right size, small enough for personal contact, big enough to take on major things, and Gold Fields have been fantastic about providing finance when necessary.

This is a long-service company. I have been with it since 1946. All our Vice-Presidents have been here between twenty-five to forty years – that's typical. Our Financial Controller, from Texas, is regarded as a new boy – only here nine years!

Johndrow showed me around the plant, originally a forty-acre site, added to in 1964, again in 1982. We went though the various shops, including the main machine shop:

We have all the latest machine-tools here, thanks to Gold Fields, which has financed state-of-the-art investment all through. We make a complete line of rigs, all sizes, for water-drilling, minerals exploration and oil exploration. The oil business is cyclical, water constant. We learned from the 1957 crisis to serve as wide a spread of industries as possible. For instance, we're making a small rig used by US Army engineers for soil-testing. Again the EPA now monitors anything stored underground. This is a tremendous new market for us, producing monitoring drills. We also make drills for geothermal air-conditioning pumps. Some drills can do both. Many of our rigs are multi-purpose. We began in 1931–32 with the first mobile rig, with a powered unit. This was George Failing's first product – quite new in those days. Since then we have made over 5,000 rigs. We sell all over the world. Turkey has 125 of our rigs, India 250, Argentina about 100, Brazil 45, Tunisia and Algeria 35 each, Saudi Arabia 30 to 35, Mexico about 150. Our profits are highly dependent on servicing and selling spare parts. So we know exactly what we have sold and to whom from the beginning. Our filing system – all on microfilm now – gives the precise details of every piece of machinery we have ever made going back to 1931–32. This is very useful for sending spares. We have competitors in Italy, the UK, France, Germany and Spain, and some of our competitors would like to get into bed with us. Sure: we are being hit by the strong dollar. Mexico has twenty-four of our $400,000 rigs. It can't get exchange control permission to buy spares from us so it's cannibalising.

The company is perhaps best known for its famous vibrator-trucks. This amazing vehicle, made in different versions for use in every kind of climate and conditions all over the world, applies geophysical principles to the mapping of sub-surface strata, and so to oil exploration. It embodies the Vibroseis system, which sends hydraulically-activated and electronically-controlled vibrations into the earth and records and correlates the resulting reflections. Continental Oil (Conoco) developed this vibro-system and in 1954 licensed Failing to develop the hardware, charging a license-fee and royalties. For a long time Failing had a monopoly of the trucks. Then Conoco decided a competitor was necessary and licensed one. Johndrow says: 'Our competitor is a very good fellow but he has got into quite a bit of trouble as he has only one product.' As the trucks are used almost exclusively in petroleum exploration their sales are cyclical. Johndrow told me: 'Historically we have built an average of about 100 to 140 vibrator-trucks a year. In 1981 we made between 125 and 130. During the first three months of 1982 we sold forty. Between the beginning of April and November we could not sell a single one. In December we sold six.' They are now selling hard to China and Russia, less subject to the cyclical trends of the West. While I was there a big order from China was in its final stages.

There were many different types in the truck shop. The vibrators come in several sizes and can be mounted in standard trucks or off-highway special vehicles. For arctic work, for instance, they are mounted on Cats with special cabins and steel screens protecting the machinery. The vibrating unit outlasts the trucks. Some go back thirty years to the beginning, and are still working. A lot of the Company's work consists of boosting or rehabilitating oil vibrators – there were four in the shop while I was there. The durability of the Failing product explains its 'Cadillac' accolade. And indeed at the heart of the vibrator truck is a spectacular piece of machinery. This is the Reaction Mass, an enormous block of special steel weighing 8,000 pounds. It takes a month to make and has to be maintained to tolerances of one-thousandth of an inch.

One of the strengths of Failing is that it builds a greater proportion of its products than any of its competitors. They make on average about fifty-five per cent of the contents of their finished rigs; sixty-seven per cent in some cases. They make their own mud-pumps, travelling-blocks, swivels, kellys

GEFCO vibrator, mounted on a tractor, used to generate seismic shock waves through the ground to identify geological formations.

and, not least, every kind of drill pipe, used in enormous quantities – China recently bought fifteen rigs and many hundreds of drill-pipes. This policy has made a big difference to profits and gives customers a feeling of security about the availability of spares.

George E.Failing takes pride in its quality products and its record; but profits come first. This emerges very clearly when you sit in on discussions among its senior executives. All decisions are geared to the market. It has its own R&D section but it is very much based on market demand. As Johndrow says: 'We don't conduct research into things the backroom boys would like to build but only into things where a definite market demand does or will exist.' The Product Planning Committee, highly sales-orientated, meets regularly. 'We may even have an order for the new unit before we start designing it,' said Johndrow; though Jim Kvasnicka, Vice-President, Engineering, added: 'Being a conservative, I like to design a prototype before we go into production.' They are very excited about the new market in monitoring drills, especially the multi-purpose one. 'It is much smaller than anything we have done,' Johndrow told me. 'It could be a world best-seller. We have engineered the technical stuff. But in developing any new product, the impulse must come from the marketing side. It would be fatal for us to take off, and say "Hey, this would sure be cute – but John McMahan [Vice-President, Marketing] can't sell it".'

Johndrow let me through the yard and pointed to one of their biggest rigs, which drills to 11,000 feet: 'It is the only one we have left unsold, I am happy to say.' It was with a certain sardonic pleasure that he drove me past the plant of one of his rivals, where we counted twenty-eight big rigs lying unsold in the grounds. 'Those guys annoyed me', he said, 'they have been trying to drum up trade with adverts offering free vacations in Hawaii.' Johndrow told me the firm had been doing $75 million of business a year at the top of the boom. Now it was down to $30 million, hence the anxiety to make a high percentage of the product, and to refurbish and update old vibrators:

> When things are like they are, you look around for things you can do. Our business consists of small businesses, and you have to be fast on your feet and get involved in as many businesses as you can. We finance a lot of transactions and we haven't had twenty-five repossessions in the whole history of the company. We want our customers to succeed. We are less interested in their finances than in their experience – we want to be sure a customer has a Chinaman's chance of surviving. Our plant can now produce annually $120–125 million of business if and when the market is right. And it will be.

George E.Failing is a jewel; but a jewel, in the opinion of Gold Fields' strategists, which was resting in the wrong crown. Consolidated Gold Fields is determined to adorn its American crown with jewels from two general areas – mining and the construction materials industry. The latter is likely to be the area of most rapid expansion since CGF has been encouraged by the confident growth and profit-record of Amey Roadstone, the big company it has put together in Britain, and would like to repeat the

Part of the ARC America Associated Sand and Gravel plant near Seattle which operates in Alaska as well as the north-west coast of the USA.

experience in America, if possible on an even bigger scale. ARC in Britain has demonstrated that Gold Fields has the skill to expand from mining into the broader business philosophy of natural resources provision. It also deals with a logical debating point liable to crop up in strategy discussions: if Gold Fields should be wary of manufacturing, which requires management skills it does not seem to possess, why should it invest heavily in the construction materials industry involving a large element of manufacturing? ARC seems to have shown that this is irrelevant when the business goes right through from the raw materials in sand, gravel and aggregates to finished products in pipes, bricks, blocks and tiles. As Bill Brown put it to me: 'Almost all successful diversification in business is following a theme or product through to the market place. In construction materials, adding value is logical – it gives a unified theme.'

That is the philosophical justification. But it must be stressed that there are two important differences between ARC in Britain and ARC America. First, the American construction materials market is colossal compared to Britain's. This means it is necessarily a regional business and, granted the vast distances, is likely to remain one to a considerable degree. Secondly,

ARC itself is a company exceptionally rich in reserves of raw materials. It has, as we shall see when we examine it in detail, a very large number of good quarries, and by comparison its manufacturing side, though healthy and go-ahead, is comparatively small. By contrast, ARC America has very little in the way of reserves but a very large production side. At present ARC America consists of four elements. Hydro Conduit is probably the largest concrete pipe producer in the world and virtually a national company serving fifteen states from twenty-seven centres. The other three are regional or even local. WMK supplies sand, gravel, ready-mixed concrete and concrete block to parts of Nevada and Arizona. Associated Sand and Gravel produces aggregates, ready mixed-concrete, pipes, pre-stressed products and asphalt for the Pacific Northwest, particularly in the state of Washington. Finally, Cement Products produces aggregates, ready-mixed concrete and block in Florida. The aim is to expand generally, but in particular to acquire a much larger reserve base. In the long run, it is hoped, both ARC and ARC America will have a much more similar balance of reserves and finished products.

The strategy behind the expansion of ARC America can be seen from four different, not necessarily conflicting, viewpoints: from London, from New York, from ARCA headquarters in Newport Beach, California and from the operating companies on the ground. In London I talked to Chris Glynn, then a CGF executive with special responsibility for the construc-tion materials side. He told me:

> From the mid-1970s the requirement has been to develop a business there akin to the UK – aggregates, hard rock, sand, gravel, pipes and blocks. We want to be in the sun-belt and the south-east. We have already spent $120 to $140 million buying things. We like to buy companies with growth in areas with population take-off, with high curve of *per capita* use of rock, gravel, etc. How do we set about it?
>
> Background papers are prepared on the aggregates industry – volumes and volumes. We also hire consultants, such as McKinsey. Then there are some contacts, such as banking contacts. We have various criteria for acquisition. For aggregates the overriding question is the reserve position. This can be very complicated since it involves not just their existence but the likelihood of getting planning permission. It also involves the existence of other nearby reserves and transportation costs. The location is vital. But there are also special factors in each case. Is a new airport likely to be built, for instance? Then there is the quality of the rock. You could sum up this way: absolute reserves; planning environ-ment; reserve position in relation to competition. McKinsey place great emphasis on what they call 'planning environment' – for instance, California is now very bad. You also look for the characteristics of the business, its market share – some pay great attention to this. Then there are also local markets, growth markets, the union position, pension liabilities. But these factors are true of any acquisition search. Secondly, on the building products side, the reserves are not important. But the anti-trust position may be vital. Here, I would say, the quality of management is most important.

In New York I discussed the divesting and reinvesting programme with Richard Secrist who, as President and CEO of GFAI, was in charge of the sell-off and, as Executive Vice-President of GFAC, was also concerned with the reallocation of funds thereby made available. He told me he saw it as follows:

There is only one quarry, in Florida. So the primary emphasis is on building up reserves of materials. In practice this means buying up companies. They tend to be very expensive. Permitting is the problem – so those who already have a permit to quarry are in a strong position and can demand a high price for selling. So it is a question of selling off the manufacturing sector to raise cash. The old style manufacturing sector was in bits and pieces, and there were no reserves of management skill to make sense of it. In aggregate manufacturing, on the other hand, there is no contradiction. It is integrated and you can switch management around. And the important thing is that we have the ability to predict changes in the construction industry. We are thinking in terms of five to ten years to complete the programme of disposal and acquisition.

At the Newport Beach HQ of ARC America, Keith Orrell-Jones, its President and Chief Executive, explained how the present spread of companies was put together.

ARC came to America in 1974 and bought fifty-one per cent of Cement Products, which was then three or four ready-mixed plants in St Petersburg, Florida. Then in 1976 came the purchase of Kyle in Atlanta – this did ready-mix, block and concrete pipe, operating in the Deep South, Georgia, Alabama and Florida. The really significant move was in 1977 with the acquisition of Hydro Conduit, which made pipes and pre-stressed concrete sections for bridges and so on, and also owned Associated Sand & Gravel in Washington State. This was a vertically integrated company: a big sand and gravel pit, producing ready-mix, pre-stressed concrete bridge-beams, piles and pipes and a contracting operation to surface roads and airfields, making its own asphalt. It operated in Alaska and the islands as well, and this means doing everything for yourself. Then in 1980 they bought WMK in Las Vegas: sand and gravel, ready-mix and concrete block. I came out here in October 1981, shortly before ARC America was transferred to Gold Fields American Corporation – though the management remains ARC's.

The orientation is product-line. But another consideration is – what is the most effective legal structure from the tax point of view? We broke up Kyle and took its pipes to create Hydro in its present larger form as a national pipe company, probably the biggest in the world. As you know, our main future growth will be in aggregates. But don't forget the big difference in the market structure to the UK. The US aggregate industry is ten times as big and the geographical spread enormous. The biggest aggregate company in America has only two per cent of the market. There are a great many small companies, most of them private.

First we decided which parts of the country to concentrate on. We adopted a sun-belt strategy because population growth drives construction – not just demographic growth but movement. Take Irvine, for instance, expanding from 7,000 to 70,000 in a decade. It creates a tremendous demand for concrete. Now: it is easy to get into the ready-mixed business. Expensive to get into aggregates. There are some markets where aggregates and ready-mixed are quite separate. Our basic approach is to go into markets in the way that market is structured. So often we have to buy into ready-mixed as well as aggregates. Back in 1981–82 we set specific targets: Washington State, for instance – we are buying a small company there today. We have a fairly substantial shopping list of companies. As Gold Fields' stock is not quoted over here, we are not able to make unfriendly bids for publicly quoted companies. But we have enough opportunities without doing that.

When we acquire a company we usually take on the Chief Executive. Here at Newport Beach we have a very small operation. The four composing companies are all self-operating. You have to remember that Chipping Sodbury (HQ of ARC in Britain) is as far from Cairo as Newport Beach is from Florida. The exception is Hydro which has a natural structure. Each of its plants has a high degree of self-administration. We shifted effective control of Hydro from Newport Beach to Houston, Texas. When we take over a company, the people who ran it continue to run it. We provide direction and financial control. There is a very high degree of decentralisation.

Our potential in raw materials, aggregates, is so great that we are less interested in the manufacturing side. So further processing is not our priority, though of course we would not turn down any good opportunity that presented itself. I have been put here to help a company grow and emulate ARC UK. That has taken fifty years to create, and you don't create another company like that in five minutes. ARC has been good for Gold Fields, and vice-versa. The effort is being made in London to associate its thinking with the rising star of ARC. In our view the local managers ought to be important if only for local environment reasons, and you have to have more devolution in America because of its size. The ARC group in the US are very pleased with ARC UK, which they look upon as their parent, rather than Gold Fields. They like it partly because the ARC people know their business.

Bill Lightcap, in charge of the financial aspects at Newport Beach, explained to me the financial links with London:

Legally our parent is Gold Fields American Corporation – for tax purposes and bank-loans. The cost of acquiring Hydro, for instance, was financed partly by institutional loan, partly by borrowing from US banks, partly by borrowing from Gold Fields, and partly self-financed. For an acquisition we turn to Gold Fields and basically it is up to Antony Hichens whether we raise the money here or get it from London. For small acquisitions – there were two in the last week – we raise money here. The profits we make here are used to finance our own expansion and capital expenditure. Periodically, Gold Fields has asked for a dividend but that is exceptional. At the moment we are in a growth

Keith Orrell-Jones, President and Chief Executive of ARC America (left), with Richard Boberschmidt and John Hutsell. Orrell-Jones takes over from Charles Spence as Chief Executive of ARC on 1 July 1987.

Transporting concrete blocks from a production plant to stockpile in Florida.

period and they are willing to see us grow. At some future stage we will return them regular cash.

Acquisitions are considered first by ARC America, then ARC in England and each of them have executive committees who see it before the full board. It is cumbersome, as bureaucratic as it sounds. For a large acquisition, the people from the Planning Department come out from England and they help to shorten the procedures. They have more experience of assessing on the aggregate side. For the pipe side we can do it ourselves. There is specific control on expenditure. At our board level (ARCA) we can approve expenditure up to $500,000 but all acquisitions have to go to England.

There are, of course, irksome aspects of strict financial control, but no one has yet devised a better method of supervising a large and diverse organisation from afar without interfering in management. You might say it is the price management have to pay for their operational freedom. Lord Weinstock has shown at GEC, among the most successful large businesses in Britain today, that a thorough financial control system permits an unusually wide measure of creative independence at unit level. Gold Fields wants to give its subsidiaries the judgemental freedom to conduct their businesses as they think best, and indeed the capital to back that judgement; but the *quid pro quo* is close scrutiny of the figures.

Indeed it is typical of the managerial independence Gold Fields gives its ARCA units that in 1981 it shifted the operational headquarters of Hydro Conduit, the most important of them, away from ARCA head office in Newport Beach to Houston, Texas, which is nearer its geographical centre of gravity. In Houston, I discussed the pros and cons of Gold Fields' methods with the Executive Vice-President, Dan Erdeljac, and four of his chief colleagues. Hydro, like so many of the best and biggest construction materials firms today, is a group formed by the amalgamation of many small family firms. In its progressive stages of expansion it has passed though the hands of many remote owners. Dick Boberschmidt, President of Hydro who still keeps his office in Newport Beach, has 'seen the company bought and sold four or five times', to quote Erdeljac. I found in Hydro's case, as I found in other instances over the world, that senior managers of good proud companies welcomed acquisition by Gold Fields as introducing a period of stability after an unpredictable era of turbulence.

It is true that the top people at Houston find the financial controls restrictive. They told me that Newport Beach, 'requires an awful lot of discussion before an acquisition agreement is reached, especially if this requires capital expenditure.' Any expenditure of $25,000 or over has to go up to the ARCA board. It takes time to go through the bureaucracy. It probably would not prevent an acquisition, they admit, but it tends to prolong inefficiency at a particular location, restricts correction of errors, and employs a very large number of man-hours. On the other hand, though specific expenditure requires approval from above, Erdeljac describes the capital budget as 'liberal'. He is emphatic that, in general, he regards the Gold Fields' period of ownership, now nearly a decade, as thoroughly satisfactory. 'They understand the business well through ARC. They are very supportive. They have provided us with the financial means for expansion.'

Dick Boberschmidt, with his long experience of other owners, put the matter thus:

It is beneficial for a company like Hydro to have an owner like Gold Fields. The financial support you get is impossible to describe. They have been very good to us in allowing us to get the right equipment. We don't have to worry. They have given us the opportunity to keep our operations first class and remain the leader in the field. This has been particularly noticeable during the recent recession. Many independents have felt it very severely and have had to cut back, whereas we have been able to keep pressing on. The (financial) control exercised by CGF is greater than what we have been used to. But we are a stronger and better company these days as a result. We used to have a different philosophy in the early days of Hydro. Then the boss didn't like managers to communicate with each other – you communicated through him. This has changed. People call each other up for information about methods of construction and so on. We now have very good lines of communication within the company. Gold Fields has encouraged this. They have encouraged us to work more independently.

As a matter of fact, however much Hydro's bosses may complain about financial controls from above, that is exactly the way they run their own thirty-five operating units. As Erdeljac puts it:

Each is structured with a plant manager provided with guidelines of accounting and financial controls. There are limits to what he can do from a dollar standpoint. Procedural items, such as payrolls, are standardised. He consolidates and issues financial statements. He reviews standard costs and criteria. Cash managements are set up under a control-disbursement banking system. They know what cash they have on a day-to-day basis. So the form of control is essentially financial – the reporting system enables them to detect something going wrong very quickly. We have productivity indices too. An acquisition is quickly standardised. Sometimes there is resistance, but those that were reluctant are not in the management slots today. So far as operations are concerned, we have certain fixed standards where they're comparable – upper and lower parameters. To some extent it depends on regional and district managers. Plant managers get a lot of latitude but they're not given a blank cheque.

Hydro Conduit makes pre-stressed beams, 'Double-Ts' and various pre-cast building components. But ninety per cent of its output is pipes. They make concrete pipes of all types and sizes, including sanitary lines, conduits and storm-drains ranging up to 144 inches in diameter. They supply under contract as a rule and sometimes sell direct to a local authority, though more usually to the contractor. They flourished mightily in the free-spending 1970s, especially when the government was giving up to eighty

RIGHT Forming steel cages to reinforce concrete pipes.

FAR RIGHT Feeding concrete into the top of a vibrating pipe mould housed underground.

BELOW Laying 54″ diameter concrete pipes at Mesa in Arizona, which were manufactured by Hydro-Conduit at their Phoenix plant.

per cent funding to Counties to improve sewage-disposal. Now public work has tapered off and they are having to fight hard for a good market share of private contracts. But they have great strength in this battle, resting partly on their ability to act with speed and on a huge scale in handling big projects, and partly on the soundness of their design, research and technological bases.

Erdeljac is particularly proud of a big Kansas City project they handled in 1979, which illustrates the Hydro spirit:

> It was worth $12.5 million, the biggest job we'd ever handled. Kansas City's Little Blue Valley sub-district wanted 33,000 feet of large diameter pipes for waste water. They took direct bids from pipe-suppliers and we won it. We built a temporary plant near the site. So we didn't have to worry about transportation weights on bridges and highways. We were in and out in under eighteen months – a very successful job both for us and for the main contractor. But, of course, this kind of Kansas City job would not be approved today. And private business is cyclical. It tends to swing with housing growth.

The key to Hydro's technology is the engineering division. This does the design and the research. It prepares complete plant-design and supervises construction. It inherits a long tradition of innovation, dating from the first concrete pipes in the United States, laid in Mohawk, New York in 1842, and designed in its office. It does not go in for spectacular new discoveries which may not work in practice but for steady, progressive improvements which promote strength, reliability, safety, durability, ease of handling and cost-savings. Dick Boberschmidt says:

> We have very high standards and quality control. Our engineering group is without doubt the best. Through this, we probably do more for the pipe industry as a whole than any other company in the country. We are promoting things like sacrificial concrete – extra cover over the reinforcing steel to lengthen the life of the pipe. This is important where they're susceptible to hydrogen-sulphide gas. We have also got a special programme for designing our reinforcing steel. This concerns the actual configuration of the rings, the placement, the combination of circular and elliptical, depending on the purpose of the pipe and its type. The design saves a lot of steel.

Hydro devotes a lot of attention to its employees down the hierarchy. 'It is a people-oriented company', Erdeljac says, 'with a tremendous amount of pride'. It puts its motto on the key-chain it distributes: 'You make the difference.' All its people are rated three times a year. There is a special stress on safety. Hydro fully accepts the philosophy of loss-control, which I have already described in relation to the mining industry; as Erdeljac says, 'We believe safety and production efficiency go hand in hand.' There is an award for the Plant of the Year: each employee gets a gift, plus a lunch. I was impressed by the Hydro plants I visited: there is nothing fancy about them, they are simple, functional, capital-intensive, cost-conscious, very tidy – important from the safety viewpoint. 'There is good housekeeping at all our plants,' as Erdeljac puts it. He stresses this as a Hydro characteristic but I would say it is to be found in nearly all Gold Fields' properties all over the world.

Hydro is a big company, ready and well poised for further expansion, though it is not going to break its neck, and make mistakes, in the anxiety to become still bigger. About ninety-nine per cent of its orders are delivered on site, so its plants operate within a small radius. Their geographical distribution is pretty even, and by their spread across the whole of the southern half of the US they cover all the main areas of growth. 'We have plants in nearly all the cities that matter,' says Boberschmidt. This is essentially a sun-belt strategy. Erdeljac says he is not really interested in expanding into the northeast. While I was in Houston they were discussing a proposed acquisition in Portland, Oregon. That made good sense: the Hydro idea is to wait for acquisitions to come along which fit into the policy. In that way growth remains natural and sound, but steady.

Natural growth, but sometimes on a hectic scale, is also the characteristic of Florida, the State of the US which has the best prospects for expansion in the near and medium term. Cement Products Corporation is basically a Florida firm and more specifically a mid-State, west-coast firm operating from Headquarters in Clearwater and covering the area of biggest potential growth. It has nineteen ready-mixed concrete plants, many of them serviced from its own big limestone quarry in the Everglades, and five block plants.

In Clearwater, Joe Devine, President of CPC, told me:

We do most of the scouting around for quarries. The negotiations are done in conjunction with ARCA. The papers are prepared by the CPC board and then put up higher. The final negotiations are done by ARCA if it's a big deal with national implications. West of the Rockies it's mainly sand and gravel. The quarrying is mainly to the east, limestone, hard rock. If we bought really big aggregate reserves we'd probably form a separate company.

We decide locally the level of capital expenditure. It's presented to ARCA and approved by them. We return profits to ARCA in the form of dividends; this is determined in relation with Gold Fields. We are a wholly owned subsidiary so all is consolidated financially at the ARCA level. I have been independently owned, privately owned and owned by three other companies, including soft-goods people, Montgomery Ward. Let me tell you it is much better to be owned by a company which understands the business, like ARC and Gold Fields. One consequence of stricter control is that it has made us much better planners. We look at proposals much more carefully. The delays are a nuisance, though. Our competitors, especially the foreign ones like Tarmac, seem to be much quicker. In the last three years Tarmac, based in Dallas, has made three major acquisitions. But sometimes these quick decisions tend to be wrong. Probably the ideal is somewhere between Tarmac's speed and ARCA's slowness. A delay can sometimes lead to the deterioration of a company you are seeking to acquire. Before Keith Orrell-Jones came we were at a standstill. Now we anticipate questions the Gold Fields' board is likely to ask. This is excellent – we are making progress.

Devine thinks ARCA as a whole 'has a very good future', Hydro Conduit in particular. CPC is now 'one of the leaders, with good prospects of expansion'. They have had their Naples quarry for three years and it has proved profitable. ARCA bought into CPC in 1974, acquiring fifty-one per cent of the stock, and under a five-year option took up the balance in April 1977. Gary Thompson, Credit Manager at CPC, told me:

Originally we had only four locations. Now we have nineteen. This is a fast moving industry. Tarmac and RMC are moving faster than we are. The established companies are moving at the same speed as us. British companies are putting their money here simply because the return on capital looks good, and I would say that fifty per cent of the Florida construction materials industry is now owned by the British. Nothing we are doing is just short-term. We are building a strong company for the future. A ready-mixed truck costs over $100,000 so you can see what the capital requirements are.

Indeed, CPC is becoming highly capital-intensive. At its Oldsmar ready-mixed concrete plant, typical of its producing units, turning out 350 yards a day from crushed limestone, there were only three men: a plant super-intendent, a materials' hauler and a retailer/plant maintenance man. Until recently eight or nine trucks were standing about there all the time. The change was made possible by the creation of what they call Central Dispatch. This computerised control-system, designed to make maximum use of CPC's fleet of 175 trucks, is the brain-child of Jim Bishop, Director of Technical Services. He told me:

The main object of the new Alcon system of computerised batching is not so much to save jobs as to eliminate human error, improve our material control and raise quality and uniformity. The two most important things, so far as our customers are concerned, are quality and service. Central Dispatch involves a central area from which all the trucks are controlled, instead of each producing company doing it separately. Our equipment is now pretty sophisticated and we are doing some things that other companies are only thinking about.

Bishop is also concerned with applying technology to safety – ARC America, like ARC in the UK, treats safety and efficiency as indivisible – and to compliance with environmental regulations. For instance, all CPC plants are being equipped with reprocessors for concrete returned from the job site. It is amazing how few building contractors can calculate volumes exactly. They tend to over-order, and the surplus must now be returned from the site and reprocessed. CPC believes, along with the best mining companies, that the only safe way to ensure compliance with regulations and avoid accusations of pollution is to achieve a closed circuit – except for reception and delivery of specific materials – and contain everything on the property. In the long run this is a formula for efficiency too, which is promoted by rational concern not only for safety but for the environment.

Certainly, an aggressively expanding company like CPC has now to plan its technical future within the framework of strict planning laws:

The environmental thing: [Joe Devine says], is beginning to be very difficult. In a way, Florida is the youngest State to modernise itself, to install a proper infrastructure, and it has a lot of land to be drained and built on. So it is a very fast growing State. The need for good specifications – in blocks, building, roads, bankings and so on – is just starting. The State transportation authorities are under pressure from the Federals to upgrade roads, bridges and everything else. This is excellent news for us. The standards required take the independent fly-by-nights out of the market and make it essential to employ highly professional firms like us. But concern for the environment is being intensified at the same time – especially as it concerns mining, quarrying, fishing, the Wetlands and so on. It's really just beginning here.

We are therefore upgrading our plants and equipment and thinking ahead. Our Naples quarry is unique in comparison with ARC's in Britain. They fill with water after they're finished, but we are mining under water. We own 1,600 acres there and we are thinking of transforming it into a saleable after-use area by digging lakes and canals so it can be developed as a residential and commercial area, with fishing and boating. In the meantime, the National Crushed Stone Association has an annual award for the best beautification of an active quarry. Our quarry is in the Everglades in the middle of nowhere and as we are working underwater we can't do much beautification, but we are dressing it up as best we can.

CPC's problems are a microcosm of ARCA's as a whole. No one who visits the United States frequently and who has visited all parts of this immense country, as I have, can fail to be impressed by the opportunities America still offers to the energetic entrepreneur with vision. In fact, after the troubled and pessimistic years of the second half of the 1970s, American capitalism experienced an invigorating recovery of spirits and confidence in the first half of the 1980s. There is little doubt that the spirit will endure for some time to come, and in many respects the United States now provides the freest and most inviting commercial climate in which to operate anywhere in the world.

But this general proposition is subject to three important qualifications. Firstly, America is one of the richest depositories of natural resources in the world. But its enterprising people have been exploiting them ruthlessly and on an ever-growing scale for two centuries. So in America has been born a determination which is not so much the product of politics, party or ideology as of native common-sense, that the exploitation process must be rationally controlled in the light of the best available scientific knowledge. Alas, in some ways this common-sensical and rational movement has been captured and perverted by fanatics. America is a country of enthusiasms, liable to carry anything, whether individualistic exploitation or collectivist restriction, to extremes. Much foolish environmental legislation was passed in the late 1960s and 1970s; regulations issued in accordance with it are often oppressive; and States and local authorities, particularly in some areas, have created what amounts to an anti-business climate. This hostility is particularly limiting to a company like Gold Fields whose business is to find, extract and add value to natural resources – large-scale and highly visible activities easily presented as anti-social.

Secondly, America is a citadel both of entrepreneurial freedom and of anti-trust legislation. It has an unbridled enthusiasm for each of them. It is probably easier to create a business in America than anywhere else, and to expand it rapidly. It is also very easy to fall foul of the anti-trust laws, which are extremely complex, and variably – though usually fiercely – applied. For a company like Gold Fields, ultimately controlled from abroad, dealing essentially in national assets, the pitfalls are endless.

Thirdly, the United States is a country under the rule of law – none more so – but it sometimes gives the impression that it is a country under the rule of lawyers. Compared with the two other leading capitalist powers, it has four times as many lawyers *per capita* as West Germany, and ten times as many as Japan. America is a litigious society, and the propensity to go to law is nowhere more freely indulged than in business. This third characteristic intensifies the risks and penalties created by the first two.

However, America *is* a country under the rule of law, and companies which make it their business to understand the law, and which take pains to oblige their executives and managers to observe it, can survive and flourish mightily. In America, wise businessmen do not waste their breath protesting about the iniquities, anomalies and restrictions either of the environmental or the anti-trust laws. They quietly study them, and learn to live with them. They learn to cope with a litigious society by having their own posses of high-grade lawyers, by having them examine the implications of each step before it is taken, and by constructing in advance sound legal defences for every action. All this is very arduous and expensive. But once such procedures become a standard and almost automatic part of company routine, and the expense is built into the company's costs, it is surprising how quickly they cease to be important or inhibiting. Gold Fields has made many mistakes in its American ventures since that first investment in 1909. But it has learned many lessons, too, and one of the lessons it has taken to heart is the need to adapt wholeheartedly to the legal framework within which business must operate in the United States. America presents a company like Gold Fields with great problems and great opportunities. It has achieved a shrewd understanding of the problems, and is now well prepared to take advantage of the opportunities. Its independent position there, however, is influenced and strengthened by its substantial holding in Newmont Mining, and we shall examine the implications of this next.

4 The Newmont Investment

DURING THE 1970s Gold Fields decided that if it was to achieve a significant presence on the American scene, and if, in particular, it was to become an important element in American mining, it would have to take a stake in a major US mining company. Natural expansion of its own mining interests was not enough. To create a US mining finance house would take too long. In 1981 the decision was reached to go for the Newmont Mining Corporation. This was a big company with a wide range of activities and interests, comparable in size to Gold Fields itself. It had been founded in 1921, going public in 1925 with a capital of $8 million. In 1971, its fiftieth anniversary, it was worth $950 million in 24 million shares, a decade later with 27 million shares the market value had grown to $1.9 billion. The decision to acquire a large interest in Newmont, if not eventually outright control, thus involved several hundred million dollars. In outright financial terms it was the most important single decision ever taken by Gold Fields. To what kind of company did CGF propose to link its fortunes?

Newmont was and is the creation of two remarkable individuals – Colonel William Boyce Thompson and Plato Malozemoff. Thompson put it together out of nothing; Malozemoff gave it its modern size and shape. There is a perennial dilemma which confronts all firms which seek to make money out of mining. Should they merely finance and promote mining ventures, and deal in mining shares? Or should they actively work the mines themselves? The South African mining finance house evolved as a happy compromise between the two. But in the United States the mining finance house system has never been established. A company is either one thing or the other. Thompson started Newmont primarily as a finance house dealing in mining properties. It gradually began to mine itself, on an ever-growing scale, and under Malozemoff it became one of the leading mining companies in America, indeed in the world, with a profound and varied range of expertise in the recovery of base and precious metals, and solid and liquid energy. It also acquired an impressive treasury of reserves. These, for Gold Fields, were the main attractions.

Not that Thompson himself was ignorant of mining. Quite the contrary. He was actually born (in 1869) in a gold camp, at Alder Gulch in Montana. But his father struck it rich and Thompson enjoyed an upper-class education, at Exeter in New Hampshire, where he formed a significant friendship with Thomas W. Lamont, later a partner in J.P. Morgan & Co. He grew up as an East Coast money-man, towards the end of the great expansionary epoch of large-scale American capitalism. But he had a nose for ore. He also discovered some brilliant people: the geologist Henry Krumb, who worked with Newmont all his life, developed the standard

Plato Malozemoff, who joined Newmont in 1945, was Chairman and Chief Executive officer from 1966 to 1985. He was the man who moulded the company to its modern shape and size.

OPPOSITE Colonel William Boyce Thompson in 1909. He founded Newmont Mining Corporation in 1921

evaluation test of porphyry copper orebodies, and left his $18 million fortune to the Columbia School of Mines; and Fred Searls, for many years in charge of exploration and new projects, who built up Newmont's fine reserve position.

Thompson was a flamboyant fellow, with a taste for the wildlife he-man adventures so fashionable in the age of Theodore Roosevelt and the Bull Moose Party. He crammed his office with hunting trophies. But his real stamping ground was Wall Street, and dearest to his heart was not big-game hunting but poker. He smoked, ate, drank and gambled voraciously. In Murray Hill, New York, he kept an apartment solely for the purpose of playing poker with his cronies, where he sometimes played continuously for a whole day and night. Not surprisingly, he died in 1930, worn out, at the age of sixty-one.

Before that, however, he had created Newmont. He set it up originally in 1916 as a company to handle his trade in securities and other bits and pieces. The 'New' was from New York, his golden city where he made his first financial killing, and the 'Mont' from Montana. Thompson was essentially

Alder Gulch in Montana where Thompson was born in 1869.

a company promoter. His method was: find potentially profitable private mining companies; form public companies to run them; sell stock to the public in these companies; then use the proceeds to build a liquid portfolio to finance further and better mining enterprises. Thompson did not merely form his own companies. He enjoyed collaborating in bigger ventures. In 1917, together with J.P. Morgan and the famous engineer Herbert Hoover, he formed part of the American Syndicate which created Ernest Oppenheimer's Anglo-American. A decade later, in 1928, he helped Oppenheimer to defeat Chester Beaty and his Rhodesian Selection Trust by taking twenty-five per cent in Rhodesian Anglo-American.

Hence Thompson was primarily a financier. But he gave Newmont a major place in the rapidly expanding copper-mining industry which was growing up to supply the electrification of the world. He was involved when the first large-scale open-cast mining of low grade copper, using steam-shovels and rail, began in 1906 near Salt Lake City in Utah. These operations created the largest man-made excavations in the world and opened a new era in copper mining. Four years later, for the sum of $130,000, he bought outright the Magma copper mine in Arizona. In due course he built a mill there, and a railway to connect it to the Southern Pacific. It proved to be a good investment, and later expanded by the acquisition of the San Manuel mine, a large porphyry copper underground mine. Half a century later Magma was still the biggest contributor to Newmont's earnings.

In the brash 1920s, Colonel Thompson's wheeling and dealing in mining stock and other commodities was part of the great American myth. The phrase 'Newmont Knows' was Wall Street gospel until the 1929 crash. But the sober fact is that, in between bouts of poker, Thompson's company had speculated wildly. The *degringolade* doubtless contributed to the Colonel's

Converting copper matte to
blister copper at the Magma
smelter and subsequently
(below) pouring copper to
form anodes of blister copper.

early death, and it left his company with only $1 million in cash and with $1.5 million due for the agreed purchase of Anglo-Rhodesian stock. In 1931–32 it made a loss, for the only time in its history. Newmont stock, which had been $236 in 1929, fell to $3.875 in 1932. Indeed the company survived at all only by borrowing from J.P.Morgan, the fruit of an old friendship.

But it did survive, and in the 1930s and 1940s re-established itself on a more solid basis as a non-speculative company actively managing a range of mining properties. During the slump it was a couple of reliable gold-mines which kept it going – exactly like Gold Fields with Robinson Deep and Simmer and Jack. Copper was no good for most of the Thirties. In 1932 its price fell to $0.09 a pound on the London Metal Exchange. Hence, though Searls had bought the rich O'okiep copper mines in South Africa in 1931, they were not re-opened until 1937. But Newmont had the Empire and North Star gold-mines in California's Grass Valley. Newmont combined the two and since gold can be sold even in the worst years, Empire-Star paid dividends despite the Depression; in February 1934, when F.D.Roosevelt raised the gold price from $20.67 to $35 an ounce, its output rose in value from $2 million to $3.5 million a year; and before it was closed in 1959 it had returned Newmont $40 million in dividends. The company also mined gold successfully in Canada.

After the racketing years of Colonel Thompson, Newmont settled into rule by oligarchy. Charles F.Ayer, a soberside lawyer, known as The Judge, was President, moving up to Company Chairman in 1947. He sat behind Thompson's enormous, superbly-carved desk, but his chief job was to keep disputes over claims out of court. Searls, a chunky bruising fellow, an accomplished amateur boxer, became President. He, in turn, succeeded as Chairman in 1954 when Malozemoff was made President. Searls built up and oversaw Newmont's properties. Operations were divided on a geographical basis. A. J.McNab, an Ontario metallurgist of Scottish origin, who designed copper smelters and lead refineries, ran Magma and made it highly profitable. A first-rate mining engineer, Henry DeWitt Smith, developed Newmont's rich, low-cost copper mines in southern Africa. Phil Kraft, a New Yorker who was both mining engineer and geologist, ran the oil properties in the US and exploration for minerals in Canada – his crowning success was the participation in the development of the biggest Algerian gas field. The real strength of Newmont has been its continued ability to enlist the loyalty of men of this calibre: men who, in addition to the highest technical qualities, have the extra-sensory instincts and entrepreneurial dash of the true mining pioneer.

In 1945 Newmont had been joined by the metallurgist Plato Malozemoff. From January 1954, when he became President, adding the Chairmanship from 1966, Newmont became much more like the autocracy Thompson had conceived, though on a more professional basis as a mining management company. Malozemoff, like Searls, had done his stint of what they called 'working the Wall Street stope' – trading in mining securities on behalf of the company. But his heart was in mining. Indeed it was in his blood. He was born in St Petersburg in 1909. His father, Alexander Malozemoff, was

A. J.McNab, an Ontario metallurgist who designed copper smelters and lead refineries, made the Magma copper mine in Arizona highly profitable.

sent to Siberia where he worked for the British gold-mining company, Lena Goldfields, in which CGF had an interest. The Bolsheviks confiscated it in 1920 and Malozemoff Senior emigrated to America. He returned to Soviet Russia in 1924 when the policy changed briefly and Lena was again open to foreign interests. Then, in 1929, Stalin seized it for good and Malozemoff returned to the US. His son Plato qualified for a Master's degree at the famous Montana School of Mines in Butte and specialised in metallurgical engineering research, though he is a man of very wide interests – he almost became a professional violinist. When he joined the company at the end of the Second World War it was on the eve of its great expansion. In 1946 its assets were less than $70 million and its earnings under $2.5 million. In 1981, thirty-five years later, with Malozemoff at the height of his glory, it was a two billion dollar company earning nearly $81 million – and nearly $200 million the year before.

The key to this great expansion was copper and the heart of it was the Magma mine. Of course Newmont, like all other major international mining companies, has been involved at one time or other in a huge range of products. It has been in oil since 1925 and its Newmont Oil Company, formed in 1944, acquired an interest in offshore oil development. More recently it has moved into coal, acquiring over thirty per cent of the stock of Peabody Holding Company Incorporated, the largest privately owned producer of coal in the world. Its interest in gold has continued through the Carlin orebody in Nevada. In 1938 it developed the Resurrection Mine in Colorado producing lead and zinc. But its big profits come from copper.

Peabody is America's largest coal producer and the biggest privately owned coal company in the world. It operates 42 mines in the United States to produce 65 million tonnes of coal a year, most of which goes to generate electricity. The dragline bucket has a capacity of over 200 tonnes and can be compared for size with the vehicle beneath it. In November 1986 Newmont acquired an additional 31 per cent interest in Peabody, which increased its total holding to 61 per cent.

Newmont has a 28.6 per cent investment in the Palabora copper mine near the Kruger National Park in north-east Transvaal, one of the world's lowest-cost copper producers.

There were, and indeed are, many copper mines in the Newmont empire. At the end of the 1930s it invested in the Flin Flon mine in Manitoba, a marvellous property which paid $52 million dividends to Newmont in its first fifteen years. The company sold Hudson Bay Mining and Smelting Company, which owned the Flin Flon mine, in 1954, thinking its best days were over; but it went on to make even more spectacular profits in the late 1950s and 1960s – the mine that was sold too soon. From the late 1930s, too, Newmont steadily developed its copper mines in southern Africa: first O'okiep in Namaqualand, then the remarkable multi-metal Tsumeb mine in South-West Africa, then the still more spectacular Palabora in the Transvaal, which Malozemoff developed in the 1950s with Val Duncan of Rio Tinto Zinc, raising over $100 million for the investment. Malozemoff was also the force behind the growth of the Sherritt Gordon mining complex in Ontario, Manitoba and Alberta, producing zinc and fertiliser as well as copper; he pursued another Canadian copper venture at Granduc in British Columbia; and created a major copper interest for Newmont in Peru, the Southern Peru Copper Corporation.

But in the 1960s, Malozemoff used the bulk of the funds generated by the rich Southern African mines to develop the Magma complex in Arizona.

His motive was partly to spread the risk away from South Africa at a time when the conventional political wisdom was that the Republic could not last. Gold Fields was doing much the same at that time. But Malozemoff's energetic expansion of Magma undoubtedly also sprang from a belief in copper. The expansion of the Magma Company and its San Manuel mine in Arizona – town site, concentrator, smelter, electrolytic refinery and rod-casting plant – was the chief source of Newmont's growth. By the early 1970s it had produced one-and-a-half billion dollars worth of copper, zinc, molybdenum, gold and silver, and represented an investment of close on $500 million. Throughout the 1970s Newmont flourished as one of the world's major copper-producers, with a wide international span but with its centre of gravity in Arizona.

This was the Megalosaurus Gold Fields began stalking from its New York office (and from London too) in the late 1970s. With Newmont then at 300 Park Avenue and GFAC at 230, the two headquarters were only the briefest of strolls apart. Surveying the decision from the mid-1980s, Antony Hichens calls it 'far-sighted and high-risk'. To buy control of or even an influential minority interest in a very big company, as big as Gold Fields itself, was a great adventure. As Gold Fields' General Manager – Finance, Roy Munro, put it to me: 'Because of the cost of the purchase, the decision to go into Newmont pre-empted many of our options. This huge investment was our big move, our major decision.' But in many respects Newmont was exactly what Gold Fields was looking for. It was not perfect; and it turned out to be much more imperfect than it seemed at the time. But then, so did many other companies. Bob Krones of GFAC pointed out to me: 'Some of the things Gold Fields looked at as alternatives to Newmont would have been disastrous. To be sure, that in itself is no argument in favour of Newmont.' Antony Hichens notes that one of the attractions of Newmont was that its characteristics resembled Gold Fields' own in some respects. First, from copper it had branched into gold, with good prospects for the latter. Second, thanks to the stake in Peabody, it was big in coal, and of good quality. Third, it was in oil and gas, including the North Sea, thus supplying something Gold Fields lacked. Fourth, it was in alloys. Fifth, it was big in copper, but not just in the US: in Canada, South Africa and Peru too. It thus had a wide, international spread of excellent reserves, but with a weighting to the US – just what Gold Fields wanted. 'The unusual feature', Hichens adds, 'is that Gold Fields went for a minority interest. The standard advice is to get fifty-one per cent, and thus control.'

Plato Malozemoff gave me his account of what followed:

In April 1981 we were advised that Gold Fields had acquired seven per cent of our stock. David Lloyd-Jacob phoned to say that the move would be filed with the Federal Trade Commission. The Gold Fields filing stated that they desired to acquire up to forty-nine per cent of the stock. So I asked David Lloyd-Jacob to come and see me to find out if Gold Fields were willing to limit its investment to twenty-five per cent. But it was then apparently not acceptable to CGF. Our lawyers advised us to oppose the acquisition. We discussed the idea of standstill and Lloyd-Jacob wished to limit that to eight to nine months. This was too short and we could get no agreement. So it began to evolve into a struggle to stop the investment, with a reference to the Federal Trade Commission.

There were various objections. One point was that there was a potential conflict of interest in the exploration field, in the United States, Australia and Canada. More important to us was the way it was done. In our view there were two ways of going into a company, by invitation or by agreement. In our case we had followed an ethic of not doing it except in the open. We had always valued the independence of the other company. We respected the desire of the other company to reinvest and not just hand out all its profits to dominant investors. Because of this we resented the Gold Fields' raid.

The principle of our opposition was that it was right that a company buying control should pay a premium, thus being fair to the existing stockholders. So we put up legal opposition. It went before the Federal District Court. The Judge said the parties should get together, instead of mucking about in his courthouse. Once the shooting really starts in the courthouse you don't get together. So I laid down conditions for an agreement. They were threefold. First, there must be a limitation of the percentage of Newmont shares Gold Fields was to hold. Second, there must be representation on the board. Third, Treasury shares should be sold directly by Newmont to Gold Fields at a premium, and this was worked out at one million shares. Finally, it was decided that this agreement, signed in 1981, which limited the Gold Fields' holding to twenty-six per cent, should extend to the end of 1984, and that a year in advance of its expiry the two Chairmen should get together to write a new treaty well in advance of the deadline.

Malozemoff showed me the joint announcement of the 1981 treaty which was framed in perspex and kept in the Newmont office. The new agreement was duly written in 1983: it raised the permissible Gold Fields' holding to 33.3 per cent and is to last till 1993. 'The deal has worked very well,' Malozemoff told me, without enthusiasm.

Hichens felt that Gold Fields' investment in Newmont was 'high risk'. The degree of risk became brutally clear when the copper price rocketed down. Newmont was and is very 'coppery'. All the same, the fact that Gold Fields signed the second treaty in 1983 shows that it intends to stick to its Newmont strategy as part of its long-term future in America. It decided to get agreement to raise its holding in the company at the end of a year in which Newmont's wholly owned copper-producing companies in the US and Canada made a combined loss of $34.6 million and two out of three of its copper interests in southern Africa also lost money. The only dividends Newmont received from its enormous investment in copper was $4.9 million from Palabora. There is no doubt that Gold Fields under-estimated the chance of a steep and prolonged fall in copper prices when it made its original investment. Hichens told me: 'Copper fell much lower than we expected – not from $1.10 to 90 cents, but to 55 cents a pound.' But having bought its stake, Gold Fields felt it had no alternative but to hang on and even extend it, in the hope and belief that Newmont's dependence on copper mining could be reduced.

The dilemma for Newmont, and to a lesser extent for Gold Fields, is that for long periods in the past the copper mines have done exceptionally well. They are usually equipped to the highest standard, employ the latest technology and are mined with great skill. The ores in the southern African mines are, or have been, very rich. O'okiep, in which Newmont now has a 40.2 per cent interest, was developed at the end of the 1930s as one of the lowest-cost and most efficient mining and smelting enterprises in the world. In its day it has been among the best operations in the industry; during the first thirty-five years of its existence, 1937–71, it paid $223 million in dividends. When the second Gold Fields–Newmont treaty was signed, O'okiep was still producing between 20,000 and 30,000 short tons of copper a year but losing money (R6.6 million in 1983–84). Placing the mine under the management of GFSA, with the object of cutting or eliminating the losses, was the first practical fruit of Gold Fields' investment and coincided with the election of Robin Plumbridge, GFSA's Chairman, to the Newmont Board.

Tsumeb, the big copper mine in Namibia (South-West Africa), was an even sadder case at recent prices. Like the Black Mountain, or more so, it is an unusually rich and complicated structure, raising exceptional difficulties and returning exceptional rewards as they were solved. Robert H. Ramsay, author of *Men and Mines of Newmont*, rightly observes: 'Few orebodies anywhere in the world have been as high-grade in as many metals, have contained so many different minerals, have presented so complex a metallurgical problem, yet have proved so profitable as the Tsumeb orebody.' Very remote, 335 miles north-east of Walvis Bay, it is, or was, a quartz outcrop, slashed by streaks of green and blue. Of the minerals in it, the most important are bornite, tennantite, chalcocite, malachite, azurite, native copper, galena, cerusite, cuprite, sphalerite, germanite, native silver and renierite. It is, in a way, a paradigm of the mineral wealth of southern Africa: for instance, only two electrolytic lead refineries in the world produce lead which is more pure than Tsumeb's. Occasionally it has reached 99.999 per cent. Incorporated in 1947, it was for a quarter-century the front runner in advance techniques of multi-metal mining. It made $1 million in its first year and by the end of its first decade was returning annual profits of $20 million. Indeed the profits from O'okiep and still more from Tsumeb transformed the financial position of Newmont in the post-war period, rather as the West Wits Line profits transformed that of Gold Fields; they enabled Newmont to think in big terms, to plan and finance its huge investment programme. Yet by the time of the Second Treaty or standstill agreement in 1983, Tsumeb was no longer profitable. Producing over 45,000 short tons of copper, nearly 25,000 tons of lead and just under two million ounces of silver, it lost nearly ten million Rand in 1983–84. However, recent improvements in Rand prices have brought the Company back into profit again.

Newmont's most valuable copper property, the only one to return dividends during the copper slump, was its 28.6 per cent interest in the vast open-pit Palabora mine and processing complex near the Kruger National Park in the North-East Transvaal. Though Palabora was created in

Tsumeb, the big copper, lead and silver mine in Namibia is 43 per cent owned by GFSA and 32.6 per cent by Newmont. Management of the mine was transferred from Newmont to GFSA in November 1986.

conjunction with Rio Tinto Zinc, which insisted on management responsibility, its development was in great part due to the enthusiasm of Plato Malozemoff. He grasped the point that its unusual structure, which has no sharp cut-off to the ore zone, would enable very deep open-pit mining to take place with a low waste to ore stripping strip ratio. He surmised Palabora would become one of the lowest-cost copper mines in the world, and he was proved right. Even at historically low prices, it mines over 100 million tons a year, mills a quarter of it, produces between 120,000 and 140,000 tons of copper, and still manages to make substantial profits. But it provides the only evidence from Newmont's properties, at the time I write, that money can still be made from copper mining.

The root of Newmont's problems lies in its relatively high-cost domestic copper mines and especially in Magma, which represents a huge investment, produces copper almost on the scale of Palabora, and loses large sums. It is also, thanks to Plato Malozemoff, a wholly owned subsidiary of Newmont. The problem, of course, is not confined to Newmont; it is

148

common to all US copper producers with large domestic output – hence the demands by the US copper industry for protection. There was a tendency, on the Gold Fields' side, to regard Newmont's belief in the long-term future of copper, including the long-term prospects of its United States properties, as emotional. This approach is ascribed above all to Malozemoff who, until his retirement was finally agreed in 1985, represented the spirit and ideology of the company, as well as controlling its actual policies. But when I discussed copper with Malozemoff, I did not find him unreasonable.

The essence of the copper problem, he told me, is:

the competition posture of companies and countries. There are very big differentials in the costs of production. The United States has fallen behind. The US copper industry used to be highly competitive. It is now high cost compared to the rest of the world. There has been a persistent recession for over three years. Prices are lower than production costs for every American company. There has been a build-up of very large stocks.

The development of metal exchanges is a very important factor in keeping the price low. They drive prices down worldwide. It is only in the last five years that the influence of traders who are neither consumers nor producers has been felt – it almost determines the prices. These factors have combined to keep prices lower for three to four years. They should have gone up in 1984 but actually went lower. Newmont, as the third largest producer in the United States, is feeling it, and it obviously disturbs our largest investor, Gold Fields.

The structure of the copper business has changed. In three centres, Chile, Zambia and Zaire, a policy of maximum production has been pursued, regardless of prices and consumption. They have been selling their copper at whatever prices they can get. That is the problem. Will consumption expand? It's true that, so far as the United States is concerned, I can't point to a trendline of higher consumption. The trendline has been horizontal for the last twelve years. But in the rest of the world there is a potential growth in consumption. It's not actually growing, but it will. The big question then is – will production continue to expand and will it produce another glut? The biggest mines are in Chile. There is at present no desire to invest in additional Chilean copper mines. But if investment does take place, there will be excess capacity. If not, there might even be a shortage. No one can say. What I will say is that I don't agree with the argument that 'copper is finished' at all. It is one of the indispensable elements in our civilisation because nothing transmits electricity or heat so well. So it is nonense to say that copper is finished.

The difficulty about the future of copper is that there is little chance the market will be allowed to function in normal fashion and slowly bring supply and demand in balance, as has happened in the case of oil supplies. The copper market has been the victim both of the growth of socialism and state ownership in the Third World, and the efforts of the West to discharge their moral responsibilities to the Third World, or as some would put it, work off their irrational feelings of guilt. Copper has been a leading theatre for the operation of the first factor, the growth of government. In the last twenty-five years government ownership of copper-mine capacity in the non-Communist world has risen from 0.2 per cent in 1960 to 43.7 per cent in 1985. Government share of actual production probably already exceeds fifty per cent. When government moves in, especially in Third World states whose economies are heavily dependent on a single commodity – copper is the outstanding example – production is determined by political factors, the need for foreign exchange being one but national prestige and the stability of the regime usually being even more important. Thus government-controlled mines ignore the market-prices and go on producing. The private sector thus becomes the swing-producer, expanding only if demand is very high, contracting sharply when demand is low. This applies at the national level, where a private sector continues to exist, and equally at the world level. Hence in an age where Third World state copper producers maintain production levels for political reasons, the American domestic copper industry has become the global swing-producer.

To make matters worse, or still more inequitable for US producers, they have had to stand by and watch while their Third World competitors were massively subsidised by the West, and particularly by their own government. Hugh Morgan, who runs the big Australian company Western Mining and is one of the few mining bosses who voices publicly what his colleagues think – and resent – in private, has made some startling calculations on this subject.

In the decade 1973–83, the International Monetary Fund and the World Bank loaned the Third World $2 billion for seventy-four commodity projects; this was in addition to a Compensatory Cash Facility designed to assist countries which depend upon exports of a few commodities and suffer from low prices. In 1982 the five Third World copper nations produced as follows: Chile 2,793 million pounds; Peru 758 million; Philippines 633 million; Zaire 934 million; Zambia 1,269 million. The last four ran their copper industries at a loss. In 1982, Morgan calculates, these five copper nations got $2 billion in aid from Western sources, Chile alone getting $1 billion. He deduces that the IMF and World Bank, to which the United States is the biggest single contributor, subsidised copper in Zambia by 9.4 cents a pound, in Chile by 12.6 cents, in Zaire by 19 cents and in Peru by 34.5 cents. To put it bluntly, Newmont, as a major tax payer, is helping to subsidise its foreign competitors whose uneconomic operations are in some danger of putting its own domestic mines out of business.

At the same time, American government legislation is raising the costs of its domestic producers still further. Malozemoff told me:

We are going through a period of studies about ways to survive in this climate. We have the largest underground mine in the world (at Magma) with very high labour costs. We think we can reduce them, but in the US the environmental restraints are very expensive. In 1988 we have deadlines to meet for compliance. It will add 18 cents a pound to our costs, which foreign producers in Third World countries don't have to pay. Ours are old, low-grade mines compared to Zaire and Chile. So the Federal authorities are in danger of killing the American copper industry. But the fact is that, if they did, there would be a world shortage. Most of my life the problem has been to find replacements for depleted orebodies. Now it is a question of how to remain in copper at all – or get out of it. There *would* be emotional problems. It has been such a large factor in our success and growth. On the other hand, Palabora is the lowest-cost copper mine in the world, and Southern Peru has great potential too. I am not optimistic about the future of copper, but I'm not prepared to be pessimistic either.

Once copper is taken out of the equation, Newmont is undoubtedly a sound long-term investment for Gold Fields. Its resources in energy, where Gold Fields, at least until the development of its South African coal mines, has traditionally been weak, are particularly attractive to the strategists in London. It has a large oil and gas potential both in the Gulf of Mexico sector of the United States and in the Dutch North Sea. Most important of all is its 30.735 per cent interest in the Peabody Holding Company Incorporated, which owns outright the Peabody Coal Company, America's largest coal producer and the biggest privately owned coal company in the world.

Newmont is increasing its oil and gas interests in the North Sea as well as American coastal waters.

The Telfer gold mine in Western Australia is 70 per cent owned by Newmont and a current expansion programme will increase annual mill capacity from 480,000 to 1.5 million tonnes a year.

Newmont's acquisition of a major interest in Peabody was not wholly unlike Gold Fields' penetration of Newmont. Plato Malozemoff told me:

Peabody was an independent company, looking for a possible merger. We were not too keen because Peabody was so big, the same size as Newmont. This was in the mid-1960s. Around 1970 they finally accepted a merger with Kennecott. But the Department of Justice objected. There was litigation. The Supreme Court upheld the government's objection. So Peabody was put out for auction at $1 billion plus. That was more than we could afford. So I organised a consortium which eventually succeeded in buying it. This deal, arranged in 1977, was quite an imaginative one – part of the money was cash, $500 million borrowed from insurance companies, $200 million was in equity and $400 million in low-interest notes over a thirty-year period. The financing was completed in 1977. We had a hard time at first. There was a strike. We recognised we had to have new, strong management. It didn't work at first. So we changed it again. Now it's good, a resounding success.

That claim is borne out by figures. Under its restructured management Peabody has achieved record earnings, despite the flat world coal market, and with thirty-six mines in ten States, an annual production of over fifty million tons and reserves of eight billion tons, it is already a substantial contributor to Newmont earnings and has the potential to become a much bigger one.

Even more attractive to Gold Fields, in an emotional sense, are Newmont's gold-mining properties and prospects. It has a seventy per cent interest in the Telfer mine in Western Australia. More significant are its interests in and around Carlin in Nevada. The discovery and successful recovery of gold by the Carlin Gold Mining Company, a wholly owned subsidiary of Newmont, was the beginning of the latest phase in American gold-mining. It dates from 1960 and was the result of the application of the principles of geophysics to gold-mining, rather like the discovery of the

The Gold Quarry development near Carlin in Nevada where Newmont Gold is rapidly increasing production.

A large diameter rotary drill used for exploration by Newmont Gold.

West Wits Line. Ralph J.Roberts, a member of the US Geological Survey, had observed from the Survey's mapping of a large area of north central Nevada that erosion had carried away so-called 'bubbles' in the very thick sedimentary and volcanic rocks which lay over the carbonate rocks beneath. This had created what Roberts called 'windows', which in some cases revealed orebodies containing mineral deposits of commercial value. A Newmont geologist, John Livermore, spotted the brief paper Roberts wrote recording his line of thought, got the backing of Malozemoff, and eventually targeted the Carlin Window as the site for an open-cast mine where low-grade ore could be cheaply recovered by milling and cyaniding. It proved to be one of the few successful open-cast gold-mines in the world at that time, and by the end of the 1960s it was producing more gold than any other gold-mine in the United States except the Homestake mine at Lead in South Dakota. It was also the first to forestall environmental restraints by developing a closed-circuit system to recycle water used in the cyanide recovery process, prevent the escape of waste and thus avoid accusations of pollution. This pioneering role in multiple aspects of gold-recovery technology reflects the range of skills Newmont possesses.

Gold production from the Carlin area has been expanded by the exploitation of the low-grade Gold Quarry orebody, raising Newmont's total gold production to half a million ounces annually, making it one of the largest free world gold producers outside South Africa and a worthy companion to Gold Fields itself. As for CGF, its interest in Newmont, added to its own gold-mines in California and New Mexico, means it now has a big stake in the new gold-mining industry of the American West.

There are two final points about Newmont's assets. First, it has a low ratio of debt to its total capital value. Despite the prolonged troubles of the copper industry, it has continued a long tradition of keeping its debt at a manageable level. Second, it retains a large portfolio of securities, deliberately kept highly liquid, providing it with the means to finance exploration and acquisitions. If Newmont is eventually persuaded or obliged to slim its copper activities drastically, as many at Gold Fields undoubtedly wish, it will emerge a smaller but in all probability a richer company.

At Gold Fields, both in New York and London, there are three different attitudes towards the Gold Fields' investment. One school of thought believes it should never have been made at all. A second, perhaps the majority, believe the investment was sound in principle but was made at the wrong time and that if Gold Fields had waited it could have acquired the same interest for not much more than half the money. In a sense this is undoubtedly true. But then – what investment could not have been better timed? If Gold Fields had delayed, Newmont might not have been available at all. It is a curious fact, or at any rate a belief entertained at Gold Fields' head office, that when its negotiators first went to the Newmont offices to discuss their initial share purchase, which then stood at seven per cent, Mr Harry Oppenheimer of Anglo-American (which already held twenty-nine per cent of CGF) was quietly in the next office negotiating to buy twenty-five per cent of Newmont. Be that as it may, there is certainly a third school of thought at Gold Fields which takes the view that, in the long run, the price paid for the Newmont interest will seem of little importance: the investment is fundamentally sound and eventually its wisdom will be demonstrated beyond question.

In the meantime, the sensible and obvious thing is for the two companies to work together for their mutual benefit. It is part of the human nature of big business that companies with long records and proud reputations do not find it easy to observe the commandment: do as you would be done by. Newmont has often bought into and acquired influential shareholdings in other companies; it says it did so only by agreement or invitation but this is

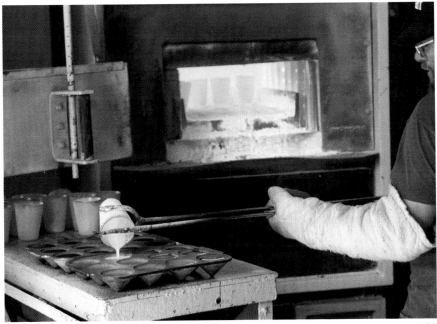

Pouring gold samples in the laboratory at Carlin in Nevada. Before the end of the 1980s Newmont Gold is expecting to produce more than 20 tonnes of gold annually.

usually an *ex post facto* rationalisation of a more edgy series of events. Yet it resents a similar move by Gold Fields. Gold Fields, equally, resented a major acquisition of its shares by Anglo-American, while pursuing the same policy towards Newmont.

When I put this point to Malozemoff, he refused to admit any inconsistency. He told me:

> Our general philosophy is – we don't want to be a minority shareholder in a company we don't like. We always want a place on the boards and a voice. In some cases this becomes a dominant voice. Excellent relationships have developed, for instance, with Sherritt Gordon in Canada. We believe in persuasion, not pounding on the table. For instance in our joint venture with RTZ, there was their phosphates proposal. We feared the phosphates market was inflated and would not last. We did a lot of studies. We persuaded RTZ not to do it, and we were [proved] right. It is important for both sides to respect each other's judgement. It's best to have a specific treaty, as we do with our joint company in Peru. There have been times when we have swung the whole company to our point of view though holding only ten per cent of the stock. With smaller shareholdings, for instance our four-and-a-half per cent in Continental Oil, we had one or two people on the board. But under the monopoly legislation this was ruled by the Feds to be a case of 'interlocking directorships'. We had to leave the Continental Board. It was the same in Texas Gulf Corporation: with time, as the companies grew, our influence grew less. In the past most American companies hated a minority position. But our philosophy has always been that a minority position in a very good thing is better than a majority position in a not-so-good thing.

Plato Malozemoff, not unnaturally, has never felt cordial towards the Gold Fields' penetration. Gold Fields felt from the start that it was likely that better relations could be established with his successors. The Newmont case fits in well with the general Gold Fields' philosophy of how to handle associated companies and their managements. Antony Hichens put it to me as follows:

> On the whole, the centre should not interfere much with the periphery, so long as they recognise you have the right of veto – especially when something is done against the interests of local shareholders as well as control shareholders.

Rudolph Agnew himself is under no illusions about the size of the Newmont problem for Gold Fields and the need for great diplomatic skill in resolving it. It is the kind of job he enjoys tackling, fitting in as it does with his notions of how the Gold Fields' commonwealth of companies should work. Certainly he is confident about the eventual outcome while by no means sure of all the intermediate steps. He told me:

> Newmont will be in the end a very sound investment and a domestic vehicle in the United States for growth in the future. On the control of

Newmont it is not easy for me to be frank with you. Technically we have twenty-six per cent of the shares, we have proportional membership on the board and membership of key committees. We don't have control to the point where we can insist they invest in X or sell out of Y. But if we came up with a brilliant idea, in the interests of all Newmont shareholders, then we effectively would have our way. It all comes back to ideas.

I admit we are not clear about Newmont at the moment. The investment was based on the belief that our hundred per cent-owned industrial activities in the United States would give us a huge cash flow during the period of investment in Newmont. The Newmont performance is much worse than we had predicted because of the phenomenally low copper price. But the real miscalculation arose from the collapse of our hundred per cent-owned industrial activities. We are now moving towards a strategy for Newmont, under its new Chief Executive, Gordon Parker, who took over as Chairman on the 1 January 1986. Eventually the investment in Newmont will look cheap. This is firstly because of the advantageous dollar exchange rate at which we made it. The assets have since sharply increased in sterling terms. Secondly its coal, oil, gas and gold assets are very good. The one disadvantage is the collapse of the copper price. The strategy, therefore, is to heal the copper haemorrhage and move the company into growth in its good assets. What it's all about at the end of the day is to have twenty-six per cent in a company which has doubled in size. Then the investment will look good. What we don't know at the moment is how to get there.

That is Gold Fields' debate over Newmont. As it happens, it is not a bad preparation for considering the problems and opportunities which face Gold Fields in Australia.

5 Building for the Future in Australia

No BRITISH MINING COMPANY with worldwide ambitions, and least of all a company with a primary interest in gold, can afford to overlook Australia. It was from the beginning and still is an explorer's country. It has a massive 5.3 per cent of the world's total surface area and only 0.39 per cent of its population. People will tell you that, from a minerological point of view, it has already been thoroughly investigated. That is not true, though the process is accelerating, as we shall see. What is unquestionable is that Australia has an immense panoply of natural resources. It may not be quite as rich as South Africa, but of the world's identified reserves it has eighteen per cent of the bauxite, seventeen per cent of the zinc, thirty-seven per cent of the zircon, seventeen per cent of the uranium, thirteen per cent of the lead, ten per cent of the black coal, nine per cent of the manganese, five per cent of the iron ore and plenty of nickel, copper and tin. It has an enviable seventeen per cent of the world's future low-cost energy resources. Moreover, Australia, though not without its political handicaps, is a mature and stable democracy, with British legal traditions, high levels of education and technology and a free-enterprise spirit. If Cecil Rhodes had gone to recoup his health there instead of South Africa, one can be quite certain he would have flourished.

All things considered, then, Gold Fields was a little slow to move into Australia, probably because in the decade before 1914 John Hays Hammond was so keen it should establish itself in the United States. Not until 1926 did CGF acquire its first major interest in Australia, a share in the Wiluna Mine in Western Australia. Its first ventures, however, were fortunate. It followed the Wiluna purchase by buying into a Kalgoorlie gold-mine, the Lake View and Star, and a one-third interest in Gold Mines of Australia Limited; in 1932 these holdings were brought together as the Gold Fields Australian Development Company. The shares were bought very cheaply, thanks to the foresight of John Agnew, a Kalgoorlie veteran, and the mines returned useful profits after the gold price went up in 1934. Gold Fields was also both active and audacious during the 1930s in pursuing gold in Papua New Guinea.

These pre-war endeavours, however, did not lead to the creation, as ideally they should have done, of a Gold Fields' mining finance house in Australia, pursuing a long-term strategy of exploration and investment. Though Gold Fields has now been investing in Australia for sixty years, it found it extremely hard to devise a company organisation there which would give it clear objectives and steady, self-sustaining growth. Only recently does it seem to have discovered the right formula. It is not easy to find out exactly why the Australian operations failed to mature; in particular, why the pre-war success in gold was not followed up, and why post-war operations, which at one time looked very promising, faltered in the 1970s. The truth I believe is that, like so many otherwise inexplicable phenomena in business, it was a question of management: not so much poor management, as management without clearly defined objectives and, above all, a sense of identity, the pre-condition of successful planning.

Rudolph Agnew, certainly, sees lack of clear objectives as the explanation of what went wrong in the past, and the key to the future. He told me:

Australia is a good example of the problems that led me to refine the confederation system as a vehicle for growth. Initially, in the late 1950s, when we had one hundred per cent ownership, the objective laid down by Harvie-Watt, to grow by acquisition, was quite clear. The plan was brilliantly carried out by Donald McCall and Gerry Mortimer to the stage in the mid-1960s at which we had established ourselves as a major force in the Australian mining industry. By then we had sold thirty per cent to the public, established a presence in Sydney, and had got into a classic business muddle. Management wasn't sure whether to be a growth or an income company, and was equally unsure who was its master – the seventy per cent owners in London or the thirty per cent in Australia. Matters were further complicated by the fact that most of the operations were themselves public companies, thus creating yet more conflicts. These conflicts prohibited vital structural changes and the company declined dramatically throughout the 1970s. Now we have completed the reconstruction. Achieving our financial targets will be a long haul but we are clear how it is to be done.

In 1980 we recruited Max Roberts as the new Chairman of CGFA to succeed Sidney Segal who had bravely taken the job on as caretaker through much of the most troubled times. Roberts, a South Australian, had just retired after a distinguished career in the oil industry and service in the United States, Britain and Australia. He was quick to see the structural flaws in the organisation. He also grasped the absolute necessity of working closely with the principal shareholder to create a better structure. The steps that followed were the product of many minds, in particular George Guise, an Executive Director of CGF, and our merchant bank advisers Schroders, but the driving force was Max.

Renison Goldfields Consolidated Limited, established in 1981, with the Australian public holding fifty-one per cent, is formed of four units: Consolidated Gold Fields Australia Limited, the old London controlled company; Renison Limited, which owns the tin mine in Tasmania; The Mount Lyell Mining and Railway Company Limited, also from Tasmania, the great old-timer operation which once ran the best copper mine in the world; and Associated Minerals Consolidated Limited, which runs the mineral sand interests. The new company has naturalised status under the foreign investments guidelines of the Australian Federal Government, a vital point. Let us now look at the new company's strengths, weaknesses – and opportunities.

What Gold Fields in London would most like to see, of course, is Renison Goldfields establishing itself as a leading gold producer in the Australian

BULOLO CAMP—

area. As Michael Lynch, the manager responsible for Liaison with Australia, puts it: 'Australian gold-mines are now small and relatively low cost. They don't require fifteen years to build.' In the past, Australia has had a very mixed career as a gold producer. To be sure, there have been some spectacular discoveries, many of them by pure accident. At Moligal in the State of Victoria, in 1869, a baker's van ran over a gutter and exposed an odd-looking piece of metal. Two miners from Cornwall, John Deason and Richard Oats, dug it up. It proved to be the 'Welcome Stranger' nugget of gold, the largest ever found, which weighed 2,248 ounces (nearly 70 kilograms) and was eventually auctioned for the then enormous sum of £9,436. Comparable nuggets turned up elsewhere. These spectacular finds provoked desperate 'rushes', which enriched the few, disappointed the vast majority, and usually petered out in a few years. As in California, there was no Cecil Rhodes to order and rationalise the confusion. No mining finance system emerged, to organise the investment of large funds to explore systematically the geology, metallurgy and engineering of gold-mining. By comparison with South Africa, the industry tended to lack true professionalism, to go for quick profits and to end by scratching the surface. In the 1920s, as Geoffrey Blainey, Australia's leading historian, has put it, 'Goldmining became an old man's industry and therefore conservative.' Kalgoorlie, its centre, was very slow to adopt the flotation process, for instance. Not until 1930 did the first flotation plant, at the Lake View mine, begin producing. The belated introduction of new technology raised Australian gold production from about 400,000 ounces a year in 1929 to over 1,600,000 ounces by the end of 1930s. With the war this boom subsided.

There was one exception to this uninspiring tale: the search for gold in the inaccessible tropical rain-forest of Papua New Guinea, a story of farsighted and tenacious endeavour in which Gold Fields was first involved

- 1932

over half a century ago. There is nothing new about looking for gold on this remote island. Indeed, it is a curious fact that wherever present day prospectors, using all the resources of modern technology, find gold, they nearly always discover that more primitive gold hunters have been there before them. The first European explorers in the East Indies knew there was gold in New Guinea. They thought it was the Biblical Ophir and they called it Isla del Oro. For unknown centuries the native Papuans have been recovering small quantities of gold from the bed of the Bulolo River and from stranded river terraces it has carved out. In 1884–85, the island was divided between Britain and Germany and towards the end of the decade a gold-rush followed. Some 27,600 ounces were recovered in the years 1888–98. It was produced by washing river gravel in primitive sluice-boxes.

In 1921 Australia got the old German colony as a League of Nations Mandate and it was one of its new administrative officers who realised that about forty million cubic yards of gold-bearing gravel could be profitably worked by dredging the Bulolo. A Vancouver company called Placer Development, in which Gold Fields had an interest, had the equipment. The trouble was getting it from the port of Lae, on the north coast, across the jungle-covered mountains, with the world's highest rainfall, to the Bulolo terraces. The problem was solved by air-power and led to the development of the first regular air-freight service. The first flight from Lae to Wau on the Bulolo was made in 1927 by 'Pard' Mustar in a De Havilland 37, with two passengers sitting in an open cockpit. The primitive airstrip at Wau was only 800 yards long and had a one-in-twelve slope. It was less than 75 yards wide at the steep end. Mustar, who called it 'the world's worst aerodrome', almost drove the plane straight into the ground and was forced to turn it head-on at the top. But he was back to base in an hour and in seven days he made ten trips carrying 2,850 pounds of freight and fifteen passengers.

Gold Fields helped to form Bulolo Gold Dredging Limited in 1930, built an aerodrome at Bulolo and supplied the bulk of the funds to form Guinea Airways. It ran three Junkers G–31 three-engined aircraft, which did the trip in thirty minutes, made ten trips a day and carried 7,000 pounds on each. They flew in the parts of four dredgers and two hydroelectric plants, all of which were assembled on the spot. Between 1932 and the beginning of the Japanese occupation, BGD was the most successful gold-dredging operation ever conducted and made profits of over $32 million. It was started up again after the war and by the time it closed down its dredging activities in 1965 it had produced about $60 million worth of gold.

Moreover, Bulolo itself was not the only stretch of the river yielding alluvial gold. There were deposits higher up at Wau, and in 1929 New Guinea Goldfields had been formed to work these upper stretches and their creeks. Gold Fields looked at these prospects in the 1960s and at first took a negative view of Wau; it looked again in 1976–79 when the gold price was rising. Finally in 1981 Renison Goldfields took over NGG and began the active modernisation and development of the Wau operation.

Wau is about 4,500 feet in altitude and in many ways has a beautiful climate. The tropical rain forest clings to the mountains but the river valley bottom is open and used for coffee plantations and cattle ranches. Despite its remoteness, many Australians like working there. RGC is understandably pleased with its Wau investment. Roger Shakesby, General Manager in charge of exploration at RGC's Sydney headquarters, told me:

It was on a hand-to-mouth basis. There was no exploration. Since then we've done a lot of exploration and found enough reserves to justify a new mill. At the moment the reserve position gives it an eight-year life, but we are sure to do better than that. We have a one-third interest in the Porgera deposits in the Western Highlands. These appear to be very large – reserves of over fifty-seven million tons at 3.76 grammes per ton. We are still at the feasibility stage there. Our other explorations in Papua New Guinea are at Kainantu – early stages yet.

Dick Patterson, RGC's General Manager Operations, emphasises that buying NGG was essentially a long-term plan:

We bought the company because we recognised the gold exploration potential of the Wau area. The strategy was to get into the area by buying NGG and then fund exploration costs from its production. The objective was, first, to explore the reserves of the current mining operation, second, to look at the reserves of the surrounding area and third, to look at Papua New Guinea as a whole.

As we have proved additional reserves at Wau itself we have put in a new mill and carbon-in-pulp plant. This will produce 25,000 ounces of gold a year, plus silver. The operation is two-fold. The major part is our open cast mine. But there's also sluicing of alluvials. Some of this is done by the locals, who pay tribute to us and we sell the gold for them. NGG's trading operations – tyres and batteries with branches in most of the major centres – brings in a very useful profit too.

Bulolo was famous in the 1930s as an example of what aircraft could do for mining engineers in remote areas. ABOVE: unloading 800 tonnes of machinery transported by plane, 31 July, 1931. BELOW: Bulolo was one of the most profitable alluvial gold-mining operations ever undertaken.

Papua New Guinea, of course, is now an independent country. It is much less edgy about 'neo-colonialism' than most new Third World countries. As Peter Fells, one of Gold Fields' London-based directors, puts it: 'Papua had the amazing good fortune of being de-colonised much later than almost anyone else.' Its government has no intention of doing silly, emotional things which drive away foreign investment. On the other hand, it sensibly intends to nurse its resources and prevent any rip-offs. Because of the remoteness and extreme conditions, investment in Papua New Guinea is horrifyingly expensive and firms are tempted to realise quick profits to ease their cash flow. In 1985 there was a serious clash of interests between the government and the OK Tedi copper–gold mine on the Fly River, 1,219 metres high on Mount Fubilan. A consortium of Amoco, Broken Hill Proprietary, Metallgesellschaft and the PNG government itself has spent a billion dollars developing this resource, which is in the bottom (fifteen per cent) cost bracket of world copper mines but high on the development cost curve because of the terrain. When copper prices fell and remained low, the government feared that the consortium would slow down development of the copper complex (including a hydro scheme and mill) and instead simply take off the mine's gold 'cap', estimated at 171 tons; so it threatened to close the mine altogether.

RGC has been careful to avoid similar clashes. Though its senior management at Wau is mainly Australian, all its mine-captains are Papua New Guineans and it is pushing training schemes. It has employed Natural Systems Research Proprietary Limited to carry out extensive surveys and interviews among villagers to discover the effect of mining, both modern and traditional, on fishing, rivers and water-resources, and social patterns; environmental plans have been produced. RGC has no intention of jeopardising its long-term future in Papua New Guinea, which in terms of reserves is promising.

RGC's area of actual and potential exploration is a wide one covering most of the South Pacific. Max Roberts, chairman of RGC, put it to me this way: 'We are exploring in the South Pacific from the Philippines down. But what I say to our exploration people is – don't use a shotgun, use a rifle. We are rifling in on Papua New Guinea.'

In the long run, however, RGC's best gold prospects may lie in Australia itself. Of thirty-six exploration projects currently being mounted by the company, twenty-five are in Australia. In particular it is looking hard in Queensland and in the Yilgara Block in Western Australia. It is also drilling at Peak Hill in New South Wales. At Pine Creek, south of Darwin in the Northern Territories, it has found and developed a viable mine: small but good. Gold was found here as long ago as 1869 though there are no records of its being mined before 1894, when it was worked by Chinese. It was also worked, intermittently and on a small scale, in the 1960s and 1970s. RGC sampled it in 1980 and diamond-drilled it in 1981. Then an operating company was formed with RGC holding sixty per cent and Enterprise Goldmines forty per cent; and production began in September 1985, with treatment due to rise to a full capacity of one million tons of oxidised ore a year, yielding 1,750 kilos of gold and 350 kilos of silver. The terrain is typical

Northern Territories: red dust, scrub, very hot, very humid. Dick Patterson is reasonably confident about the financial success of this mine: 'We are the operators. We expect to produce about 53,000 ounces of gold a year. There are reserves for a nine-year life and further exploration is going on.'

According to Shakesby, 'gold is now the dominant target' of RGC's overall exploration effort. 'We have about fifty geologists, plus those engaged in actual mining operations. They work from offices in Perth, Burnie on the north coast of Tasmania, Canberra – a large office because it handles South-East Australia and has our research and services group – Brisbane, Lae in Papua New Guinea and Manila. We are spending $(A)10 million a year, about $2 million in Papua, $1.5 million in the Philippines and the rest in Australia. Most of it goes on gold.'

RGC's exploration programme is conducted at the very frontiers of technology. This is inevitable, given the size of the areas to be covered and the harsh physical conditions. The whole art and science of searching for metals is going through a revolution. One Australian mining authority, Sir Russel Madigan, Deputy Chairman of Conzinc Riotinto Australia (CRA), had listed some important new areas in which exploration is changing dramatically: global tectonics, the study of deep earth developments; remote sensing image enhancement; airborne radiometrics; by these means details of the geological formation pattern can be discerned beneath the soil cover. These new techniques are accompanied by further advances in the more traditional methods of geophysics, geochemistry, prospecting, drilling and, not least, vastly improved maps of geological and geophysical features and other regional records. Madigan thinks 'the explorer of the future is going to find more orebodies while sitting at his desk than by tramping through the Spinifex.'

This may well prove true: when I visited RGC's exploration headquarters at Perth, covering Western Australia – probably the most important sector – I found that all its geologists were back in the office for the latest photo-interpretation course. Hilmer Geissler, their boss, told me:

We also use satellite pictures for interpretation. There are some very sophisticated methods now. You can scan the earth's surface for all sorts of elements. You don't initially need a geologist driving around the bush. He only goes in after the scans to pick up an anomaly. This is the theme of the future, though our operation is too small to cover the full range of developments: at the moment we would hire contractors to do that sort of survey.

Here we have some of the oldest rocks in the world, like South Africa. The rocks are deeply weathered, sometimes up to 100 metres, so all minerals are oxidised and hard to detect on the surface. Hence we mainly use geochemistry, that is taking soil/rock samples and analysing them, for our exploration work. The electrical geophysical methods have not been very successful here. It works well in Canada where the rocks are a little weathered. But the problem in Western Australia is that the overburden is very saline which makes electrical patterns very hard to interpret. With geochemistry, on the other hand, even if the ores are leached out in the overburden, soil and rock samples will reveal anomalies which, in turn, reveal the ores below. So you know where to drill.

We probably spend thirty per cent of our budget on drilling. There are two methods we use. One is percussion drilling which hammers a hole into the rock and gives you your sample as chips, blown up from the bottom of the hole. Then there is diamond drilling which cuts a cylindrical core from the rock. This is two to three times more expensive, but you then have larger, more informative samples and you know exactly where they came from.

We set about it like this. The whole state is covered by maps from the Geological Survey – 1:250,000. If you are looking for a particular type of orebody, the geologist will pick out a prospective area in accordance with his interpretation of the maps. Then he will go there and have a look, and if he sees likely rock he will get samples and analyse them. Here we are looking for gold, which takes up to eighty per cent of our effort, though we are interested in base metals, copper, lead, zinc, as well. If, after his visit and analysis, he feels there is an anomaly suggesting orebodies, we draw up a programme which includes acquisition of exploration and prospecting licences, soil sampling and magnetics and we then, after this, take another look at the prospects. If this confirms the anomaly, we go out drilling. We start with percussion, then if it's promising enough we go on to diamond – or if it's a deep hole on to diamond straight away.

One critical aspect is mineral tenements. You can only extract minerals on properties you have title to. So you have to take out a mineral tenement – that is, an exploration licence or a prospecting licence. The first is for a maximum of two hundred square kilometres and if you take out the full amount you must spend a minimum of $(A)60,000 for five years. A prospecting licence is much smaller in area – two square miles (or multiple) with an expenditure commitment of $(A)8,000 a year.

An EL or a PL does not give you an automatic mining right. You have to apply for a mining lease, valid for twenty-one years, before you can start mining. The expenditure commitment on a full mining lease is $(A)100,000 on 1,000 hectares. The minerals belong to the Crown throughout Australia. Here, you don't have many problems with landowners outside the south-west wheatbelt – most leaseholders further north are pastoralists. But wheat farmers are freeholders and mining firms must come to a compensation agreement with them. This can be very difficult. The new mining law of 1981 has made the State government, the miners and the farmers unhappy. In effect, the farmer has a veto. This may be changed. The Aboriginals have areas of reserves and they have a tribal veto. Up to the 1981 law you could peg the ground and sit on it without doing anything, just waiting for a big offer from a mining company. That has been made more difficult. It is usually too expensive to buy freeholds just for exploration. However, the geologically interesting areas are not usually in the wheatbelt.

Gold Fields has been active in Western Australia since the late 1920s. In those days it was gold. Then, in the late 1960s up to the mid-1970s it was mainly nickel and base metals. Outside the mineral sands developments and the Eneabba coal discovery they were not successful. The coal is being kept in reserve. It's low quality coal and there is no demand at present. But if an aluminium smelter or something similar is built north of Perth, Eneabba coal will come on-stream.

Farm-in or joint ventures are quite a popular form of exploration. A little chap or even a large company may not wish to spend big sums on development, so they take thirty per cent, say, and the rest is done by a big mining company. Goldsworthy Iron is a case in point – CGFA did not find the deposit but farmed-in. Individuals who make interesting discoveries will tend to go to the company with the best reputation – that's one reason it's important to keep up the effort.

Now the main emphasis is on gold again. In the last five years, Western Australia has proved the best for gold discoveries. About eighty per cent of the output now comes from here – twenty-three metric tons in 1983. There are about twenty new gold projects under construction or at the feasibility or final exploration stage. The projected gold production for 1987–88 is forty-six metric tons. Luckily there is as yet no tax on gold produced in Australia. You pay an application fee and rent per square kilometre, but no royalty. So a gold-mining company only pays payroll tax. The historical reason is that during World War II gold could only be sold to government. They were subsidising mines to find gold and cart it to State stamp batteries. The State refined the gold. Now, since the mid-1970s there is a free market in gold and subsidies have stopped. There has been a lot of talk about taxing gold, but this would bankrupt many smaller operations. In Western Australia they are often small. So the government would not actually gain anything and would lose in payroll tax and unemployment pay.

Government charges royalties on base metals at different rates, and taxes company profits as well. In any case, most base metals are out of favour at present owing to price-falls. Uranium? 'Uranium is no good, owing to political inhibitions' – Australia has become positively hysterical about anything to do with nuclear production. Diamonds? Geissler showed me a photograph of CRA's phenomenal diamond site in the Kimberley Mountains. 'Diamond exploration is very costly and there are problems of marketing due to the De Beers monopoly – so unless you come up with a really impressive strike like CRA's, it's not worth it.'

An exploration effort is at the mercy of government laws, fashion, and above all company enthusiasm. Geissler told me:

The Northern Territory is in a terrible mess over land claims. At present there are 3–4,000 geologists in Australia without a job. A few years ago they were flying them in from Europe. It's a very volatile business. RGC is one of the best companies in its attitude towards exploration – management, right to the top, has got an optimistic attitude to exploration and provides good budgets. This is excellent because now is a good time to pick up promising prospects. The oil companies, which went into mineral exploration from 1973 onwards, when they were flush

The 18-mile railway which links the Mt Lyell mine to the port at Strahan opened in 1898 and closed in 1963. It was used to bring in supplies and take copper out.

with profits, are now pulling out, especially in Western Australia. They are hardly employing any geologists at all. So it's good to be with RGC. Remember, the lead-up time to an orebody is ten years – six is the absolute minimum. So if you want to remain a mining company you must explore. Of course it is expensive – but buying-in is even more expensive.

Gold exploration was at a standstill between 1950 and 1978. Now, in the last seven years, many orebodies have been found, especially in Western Australia. But it takes perseverance and money. Western Mining is a case in point. They have been devoting forty per cent of their profits to exploration – and it has paid off. Newmont is good too. Now RGC is acquiring a good reputation and is attracting prospectors.

RGC's earnest search for gold in the west is in tragic contrast to its surplus of copper in Tasmania. It is one of the ironies of mining that you always seem to have too much of one metal and not enough of another. When Gold Fields paid £3 million in 1964 for a sixty per cent interest in The Mount Lyell Mining and Railway Company Limited in Tasmania, one of its chief motives was to get into Tasmanian tin, for in 1960 Lyell had taken a forty-nine per cent interest in the nearby Renison tin mine. But in buying Lyell, Gold Fields bought a famous and dramatic piece of Tasmanian mining history and an emotional crisis. For the Lyell mine is ultimately doomed, and as Antony Hichens puts it: 'One of the nastiest things in mining is having to kill a mine and wipe out a small town.'

Tasmania is quite unlike the rest of Australia. It has a different ecology and flora and fauna unique to itself. Much of the north-east is like the dairy country of southern England. The uplands of the centre remind me of Ross and Cromarty in northern Scotland. The west is a savage wilderness, to which there is no counterpart anywhere else in the world: a rocky coast, battered by tremendous seas, sullen mountain ranges of garish rock, vast forests cut only by foaming rapids, and thick scrub which has to be experienced to be believed – it can take a month to progress fifteen miles. The weather is howling gales, intermittent mist and often torrential rain, and fierce snows in winter. Yet this is where the minerals are. Sailing off this awe-inspiring land in 1642, Abel Tasman suspected metals lay there, noting: 'Compass erratic, varying eight points from one moment to another.' The British government picked the coast as a suitable spot for a penal colony; and it was escaped convicts, in the 1820s, who were the first explorers, led by Jimmy the Pieman, eventually hanged for murder and cannibalism. My diary records:

Met at Wynward Airport and driven to Burnie, where they load the copper. Inspected wharf with Mike Ayre, manager of the Lyell mine. Then driven to Queenstown, the copper town, by Laurie Jacobs, an old-timer and historian. First dairy farms. Then thick, sinister forest. Great gorges. Very wet – icy cold. They say it is the coldest and wettest summer they have had for twenty years.

There is plenty of wild life here. Saw badgers, porcupines. There are also fierce Tasmanian Devils, and recently someone saw a Tasmanian Tiger, though it is supposed to be extinct. Cornelius Lynch, the Irishman who first found gold here in the 1880s, believed that the native tigers always attacked at night, and he slept with a tomahawk, a bill-hook and a loaded revolver. Actually, it is the bush here which is the real enemy. It is called *anodopetalum biglandulosum*. It grows out horizontally and forms a mat many feet thick.

Rounded a corner and saw Queenstown, with its bare, battered mountains, Lyell and Owen. Charles Gould, the Victorian geologist who first prospected this area, called the peaks after his scientific colleagues Darwin, Lyell and Huxley, all theologically suspect, and their opponents Sedgwick, Owen and Jukes, staunch Christians. Alas, it was the agnostic mountains that produced the metals. The landscape is hideous. Sulphur fumes have turned the green all yellow. Prospectors came here in 1885 looking for gold and found copper. But in its first few years the mine was saved by its silver output. It did not pay a dividend till 1897, then became a bonanza. By 1905 Mount Lyell was the richest mining company in Australia.

Queenstown has about 5,000 people. It was famous for mining characters: Old Properly, 'Ot Dogs, a cockney, Nugget Reid, Hot Stuff and Jack the Savage. But the great man was the metallurgist Robert Carl Sticht, who introduced the pyritic smelting process and ran the mine with enormous success for many years. Laurie Jacobs showed me round the

ABOVE Excavating foundations for furnaces about 1896. These were used to smelt the concentrates, producing blister copper.

RIGHT The first mine office at Mt Lyell.

house Sticht built on a platform he cleared of bush, high on the hillside overlooking the town. The mine manager was the king in those days. Sticht had a chunk of a neighbouring hill removed so that, when he came out onto his bedroom balcony in the morning, he could see the smokestack of the smelter he had built and judge from the smoke whether it was working properly. The house, still occupied by Mike Ayre as manager, is a superb example of Edwardian capitalist architecture. Vast billiards room. Stained glass windows. Here Sticht kept his fine library of mining books (still here), medieval manuscripts and collection of Dürer woodcuts, now in the national museum. They still hold a big party here for the mine staff every Christmas Eve, and a special one for anybody leaving after ten years' service. People stay a long time at this mine – some are fourth generation.

Next day I went down the mine:

> Attended Superintendent's Conference at 8.30. Department heads reported progress, accidents, complaints, etc. 8.45 changed and went underground. Rain, very muddy. Down main decline to 30 Series, blasting level. Then to level below, and down to bottom of ore-pass. Atmosphere quite different to South African mines. No frills. No deference to superiors. Very dark. Deep mud, very difficult to walk in, like Western Front in Great War. Saw maintenance area. Then to bottom of production area of 40 Series, where they are preparing for the next four years' mining, all they have left. Then to Number One Shaft. Very cold. Fierce ventilation wind. Down to 17 Level. To pump station and into its 'mud-extractor', the only one of this type in Australia. It was designed and built thanks to Gold Fields' experience acquired in South Africa. Then up to surface and to mill and flotation plant. They used to smelt and refine the copper. But now all is closed and the smelting is done in Japan.

Mount Lyell is nearly a hundred years old and is the most famous mine in Australia. It has made colossal profits in good times. It has also been under periodic threat of closure. As the historian, Geoffrey Blainey, has put it, Lyell has faced and survived a crisis 'at least once in every decade'. Its first long boom ended in 1918 and it nearly shut down in the 1920s. It had another tremendous boom between 1956 and 1966 when it processed two million tons of ore a year and produced more copper than ever. Then, not long after Gold Fields acquired it, it faced a major decision. Dick Patterson, who spent ten years at Mount Lyell, ending as General Manager, described to me what happened:

> We had been working it open cut – the West Lyell Open Cut, as the basis for production for some thirty years. But that was getting too deep. Economics indicated that we should go underground or we should shut down. Working on current prices – this was 1965–66 – they decided to go underground. This is the start of the modern era of the mine. The capital expenditure was very heavy in the period 1967 up to 1972–73. The approach was two-pronged. First was to establish initial production workings by declining immediately below the open cut. Second, to sink a

shaft which, together with associated facilities, would enable longer-term production at depth. In the mid-1970s we were hit by the oil price shock and a severe slump in copper prices which caused serious financial difficulties and as a result it was decided to stop the shaft in an incomplete state. Ever since then the history has been one of struggle, with mining methods based on a relatively short-term view of the future.

There was a period in 1977–78 when we again faced closure. With Federal and State support subsidies and interest-free loans we weathered the storm. Then in the last few months of 1984 we faced a new crisis because mining was approaching an uneconomic depth and any extension of mine life after the late 1980s would have required a large investment in new deeper production facilities. It was not justified in our view. Therefore closure was inevitable. We considered two alternative closure scenarios. One was to close quickly, at the end of 1985. The other was to close four years later in 1989. We have now made arrangements with the Tasmanian Government to phase the closure over a longer period with State support. The workforce had already fallen from a peak in the region of 2,000 to 1,200 in the mid-1970s; now it is only 530. It is probably the most efficient mine in Australia. It is very large-scale, it is innovative, it has the latest equipment. But in mining low-grade ore in a high-cost country and in the face of falling copper prices the mounting financial pressures have led to a decision to shut it – the problem we have been debating is the optimum way to shut it, in industrial terms, in human terms and in financial terms.

Dick Patterson's view reflects the inevitable decision to which RGC was driven by harsh financial facts – Lyell lost over $(A)4.9 million in 1984. RGC's Chairman, Max Roberts, is gloomy about the whole future of copper. 'The trouble with copper', he told me, 'is that it is never consumed. All the copper that's ever been mined still exists. Certainly, there is no future for copper mining in the United States. People there live too high on the hog. It's different in emerging countries which produce it. There it is a "social metal" and their costs of labour are so low.' He tells a wry little story. An executive of Exxon reported nervously to the Chairman on an Alaskan North Slope drilling on which they had spent $500 million. 'Some good news and some bad news, chief. The bad news first?' 'Yes.' 'OK, Muckluk is a duster. There is no oil there, Chief. Now for the good news, Sir. I am glad to tell you that there is no copper either.'

Yet copper miners, even today, have a passionate attachment to their metal. Of the base metals, it is the only one that has beauty. Over dinner in Queenstown – delicious local lobster and chips – with some of the senior executives of Mount Lyell, I heard the case for copper put with touching eloquence. 'Not enough copper is used because too many people have got the false idea that it's an expensive metal. In fact it's the same price as aluminium.' 'Yes, and you wouldn't want to hang an aluminium ashtray on your wall, would you?' 'The trouble is, people think that copper is somehow old-fashioned. It's an antique metal, whereas aluminium is seen as a "modern" metal.' There is a great hatred of aluminium among the

copper folk. What makes the threatened closure of Mount Lyell so hard to take is that, as Patterson says, it is from every technical point of view a very good operation. Over dinner I was told: 'Last year we had a near-record output. We are producing the same amount of copper as a few years ago with less than half the workforce. The gains in productivity have been enormous.' 'Yes, and at the moment, thanks to the weak Australian dollar, we are actually making a profit.' 'This is a very go-ahead mine.'

The Australian Copper Development Association puts the case for copper equally vigorously. They refuse to accept that copper is being superseded in its basic functions. True, fibre optic cables have limited the use of copper in telecommunications; but this is only eleven per cent of the copper market. Plumbing is twenty-two per cent and the electrical industry forty to fifty per cent. Copper alloys are being preferred in engineering as opposed to steel, aluminium and plastics. It is favoured by environmentalists because of its ability to impede microbic growth. Its new uses include modular techniques in electronics and central diagnostic systems in car ignition, both of which demand connections using copper. According to the ACDA the threat from aluminium to additional copper uses is now over. Certainly, its overall use in advanced countries is not increasing. But the developing countries average only ten per cent of the *per capita* copper

Mt Lyell in full production during the first half of the century. Smelting was discontinued in 1969, thereafter concentrates were shipped to Japan for treatment.

173

Mt Lyell showing the open pit
which was worked out in the
1970s when all production
continued underground.

Loading an ore carrier
underground at Mt Lyell.

consumption of the mature economies and as they in turn mature the demand will increase. They see China as the great potential market of the future. They believe world copper consumption in the present decade will exceed 950 million tons. At present no one is looking for new copper deposits and stocks are falling. According to the Association, consumption now exceeds production by 300,000 tons a year. They argue that a copper famine is at hand.

Perhaps it is. But all the major copper producers are capable of increasing their production dramatically the moment they see a real market for it. Nor does this case answer the point that the future of copper lies in those countries with high-grade ore, low production costs and government patronage. Meanwhile, as Mike Ayre put it to me soberly, once a mine has ceased to invest in the future, it begins to run down automatically. At Mount Lyell, with every day that passes and no provision taken for future working, its chances of survival decline.

Mount Lyell is a first-class mine, but doomed. The Renison tin mine, thirty miles away, is also first-class but is able to use its formidable engineering, laboratory and management skills to stay in business. Its difficulties spring not from low-grade ore and high-costs – quite the contrary – but from the organisation of the world tin industry and its effort to regulate supply and keep prices stable.

In recent years, supply has exceeded demand and the International Tin Council sought to reduce supply by imposing production quotas on suppliers, and at the same time stabilise the market by buying and selling through the buffer stock manager in an agreed price range. However, production by suppliers outside the ITC, particularly Brazil and Bolivia, and some smuggling of concentrate in the Far East, undermined the production quotas and, by October 1985, the buffer stock manager no longer had funds to continue price support buying nor to honour many outstanding contracts. The full consequences of the subsequent confusion and resulting permanent closure of the London Metal Exchange market in tin is still not clear. Production quotas have been abandoned and the price of tin has dropped to half of its level before the débâcle. An equilibrium price will emerge from the market place, but when and what it will be it is too soon to say. All of these events were in the future when I made my visit to Renison, though it was already clear that the system was doomed.

Output restrictions were particularly hard on RGC since in the late 1960s and 1970s Gold Fields made a highly successful effort to transform it from a small producer, processing 100,000 tons of ore, to an ultra-modern major producer with a 800,000-ton throughput and huge economies of scale. As Dick Patterson told me, 'We can still make a profit at a sixty per cent quota given current prices but it's hard work and profitability is very adversely affected.' The key aspect of Renison policy under RGC ownership has been to improve the recovery rate. John Butler, RGC's Chief Metallurgist, explained to me:

Froth flotation in the tank house at Mt Lyell. Copper mineral particles are recovered on the bubbles.

Thanks to the quota, what we are doing at the moment is making a higher grade product to the detriment of quantity i.e. recovery. We are

The power house and a prospecting trench at the Renison tin mine, thirty miles from Mt Lyell, in the 1890s.

quoted at 930 tons of tin in concentrates each quarter. The government does not care about the percentage content. So we are trying to raise the current fifty-two per cent content to sixty per cent. If you raise the content percentage you reduce the recovery rate but improve the payment terms from the smelter. It means wasting resources but increasing profits. This is bad metallurgy but good business. It will stop as soon as quotas are lifted and we get back to full production. Under the quota we can see only 3,270 tons a year. We have the potential to produce 6,500 tons – and make full use of the ore.

John Mitchell, General Manager at the tin mine, who showed me around, is also opposed to quotas as destructive of efficient mining and detrimental to good labour relations. My diary records:

Mitchell explained the general running of the mine and plant. Tour of concentration plant very exhausting and dirty. Some floor areas, near fierce machines – grinders, revolving drums, bubbling vats of fluid metal – were covered in greasy, grey scum and fiendishly slippery. After

The crushing plant and concentrator at Renison was built by the Boulder Tin Mining Company in 1907.

climbing up and down gantries and slithering over iron gratings for forty-five minutes, I was finally let off. . . . The mine was bare, empty, dark, but no mud, thank God. Mitchell says: 'We used to be one of the lowest-cost producers in the world because of the high grade. But with quotas our costs of production have gone up – our fixed costs are high, so with reduced production, costs rise. It is our objective to lower the world production cost of the curve. It doesn't matter how much tin is being produced, as long as we are at the bottom of the cost-curve. The reserves are very extensive. But if the price dropped from its present $(A)15,000 a tonne to $(A)8,000, most of the reserves would be gone.' He told me that quotas have put some firms in Malaysia out of business. In Australia all firms remain but have cut production. As a manager he would prefer a free market. Quotas have meant a five-day milling week and reduced earnings for many employees. They had to shut down at Christmas for four to five weeks. He told me: 'You get industrial trouble when you reduce men's earning potential.' And it was an unsatisfactory way to run a plant, especially not knowing when quotas were coming off.

The underground operation is highly mechanised. However, the deeper you mine with declines, the more difficult it is, and costs rise.

(They have moved from twenty-ton to thirty-five-ton trucks to increase productivity.) Eventually you have to drive a shaft. A shaft site is currently being prepared.

The key to the mine is progressively increasing the efficiency of the concentrator plant. It has to redress the fact that ore grades are falling by raising the recovery percentage. It has also to balance the effects of quotas which raise production costs. It does this by a very complex method of multiple treatments, which make the plant a maze of circuits. Mitchell says, voicing a typical miner's grievance against metallurgists, that 'the plant runs seven days a week, the metallurgists work a five-day week and it always works best at the weekend when they're not fiddling with it.' True: but these continual improvements are vital to the survival of the mine. The Chief Metallurgist explained just a few of their doings to me. This is high-tech, and laboratory skills are employed in the most brutal and practical manner, by a team of five experts who work round the clock on the plant, taking bits out of service, re-circuiting the flow, improving the bits, then joining them up again. Sometimes things go wrong. Huge spillages mean that a re-designed circuit is overloading at one point. But they are very enthusiastic and reminded me of the clever young men at Black Mountain.

Examining the running of a mine like Renison Tin is a reminder of the salient fact which is central to the understanding of the mining industry. Not only is every mine different: each can be mined in many different ways. This is where the skills of the General Manager and the Underground Manager are vital; and the skill of senior miners in carrying out their plans. Moreover, the nature and quality of the ore are constantly changing and require close study and constant modification of the mining approach. Equally they require changing responses in the concentration process. Like mines, every recovery plant is different and constantly changing, if the boffins are clever and energetic. Mining is an endless battle against cost and nature, and the success of a firm like Gold Fields depends greatly not merely on their headquarters' strategists but on the skill of the fighting troops – especially the line managers and their chief technicians.

I learned two more things about mining in Tasmania. First, though all mines are born to die, and nothing is more pitiful than the death of a mine and its local community, mines do not necessarily stay dead. There can be, as it were, a transmigration of a mine's soul, so that it is born again, not necessarily in quite the same form. The little town attached to the Renison Mine, Zeehan – so called after one of Tasman's ships – has known these deaths and re-births. Once it was a mining boom-town, with 10,000 people, twenty-four hotels, a big theatre. Then it was a ghost-town. Now, albeit on a smaller scale, it is again flourishing. It even has a golf-course constructed on flattened mine-dumps. It boasts one of the best museums I have ever seen, devoted to Tasmanian mining history and equipped with its own eighty-six-year-old Old Timer, who remembers Zeehan in the 1900s and whose exploits are recounted to visitors by his admirers: 'Don't forget I've been to gaol too,' he piped. Not far away I visited Waratah, company town

OPPOSITE ABOVE A general view of Renison today. The lush vegetation is a result of heavy rainfall at the mine of between 100–150 inches a year.

OPPOSITE BELOW Drilling to determine ore reserves at the Wau gold mine in Papua New Guinea, where output has recently increased to 0.8 tonnes a year.

of the Mount Bischoff Mine. This was once the richest tin mine in the world, producing between 1882 and 1929 the then colossal quantity of 56,000 tons and, in the process, killing off the Cornish tin industry for a time. Waratah died but it has come to life again, a delightful little town bisected by a stream with bridges and waterfall and crowned by a superb mine-manager's house – a long, low bungalow whose verandahs are richly decorated in lacy white ironwork. For that matter, Cornish tin is enjoying a revival too (or was, until the collapse of the price forced an agonising re-appraisal as yet not concluded).

Waratah indicates that a mining town need not be ugly. Miners value beauty as much as anyone else. They hate to be branded as despoilers of nature. The environmentalist lobby is much hated in Tasmania. It waged a ruthless and successful campaign to prevent a new hydroelectric scheme harnessing the Franklin–Gordon river system just south of the mining areas. The people of Tasmania voted overwhelmingly in favour of the dam but the Federal government, under pressure from the lobby, vetoed it. But the miners of Tasmania are not against conservation. Far from it. John Mitchell took great pride in showing me the tailings-dams at Renison which have made it virtually a pollution-free mine. With each new tailings-dam the lessons of its predecessors have been incorporated, so that the aim of discharging pure, clear water from the site has become part of the technology of the mine. That is as it should be: it is part of Gold Field's philosophy.

But it is worth noting – and here is the second lesson – that the prime despoiler and polluter in this region (as in many another) is mother nature herself. In February 1982 and again in January 1983, Zeehan was nearly destroyed by bush fires. John Mitchell showed me the extent of the ravages. The fires simply leapt out of the ground. One expects bush fires on the dry Australian mainland, but in Tasmania they come as a surprise. Yet they are just as common: I saw their depredations everywhere. The truth is that, despite the very high annual rainfall, a bush fire is liable to occur the moment it fails to rain for three to four days. Many are actually started by people burning rubbish. My driver told me they were much more careless than they used to be and that, though such fires do more damage to the environment than all human activities put together, the environmental lobby shows little interest. There is a parallel here with accident prevention as merely one aspect of loss-control. Only when human pollution is seen as one part of the totality of events which change the environment will we move towards a rational policy for controlling them.

The point is an important one for Australia, which has a long history of irrational controls on the mining industry. One of the worst examples was the government ban on iron-ore exports. It was imposed in July 1938 partly because the Prime Minister, Joseph Lyons, did not want Japanese importers nosing around the unguarded north-west coast, where the biggest deposits lay, and partly because the government had been wrongly informed that Australia had only limited reserves of economically recoverable ore. The ban was not effectively lifted until the early 1960s, delaying the expansion of Australia's iron-ore industry for a quarter of a century. It is now one of the country's most valuable exports: in 1984 iron-ore shipments from Western Australia alone reached the record level of 91 million tons, and the reserves are bigger than ever.

When the export ban was lifted in 1961–62, Gold Fields, together with its partner Cyprus Mines (of California), bid for exploration leases and got three areas approximately 1,400 kilometres north of Perth. These were Mount Goldsworthy itself, which gave its name to the company; Shay Gap, known as 'Mining Area B', and an area in the Hamersley Range to the south, known as 'Mining Area C'. Gold Fields' holding was originally one-third, but this was increased first to forty-six per cent, and more recently to fifty-eight per cent, evidence of Gold Fields' long-term confidence in the Australian iron-ore industry and its own leases. Goldsworthy is unique among CGF's Australian interests in that it was not included in the reconstruction of Renison Gold Fields but is administered direct from London.

Goldsworthy, the creation of Gerry Mortimer of CGF and Paul Allen of Cyprus Mines, the first two General Managers, came on-stream in 1966 and was the first mine in the Pilbara region to produce ore. My diary records:

Left Perth 6.10 am with Bob Hannaford, product development manager at Goldsworthy Mining Limited. Flew first to Newman. Tropical damp heat at airport and millions of flies, so that immediately on getting off the

A jumbo drill at a face in the Renison tin mine.

Mt Goldsworthy came on stream in 1966 and was the first major mine in the Pilbara region of Western Australia to produce iron ore. Production today is from this open pit at Shay Gap.

OPPOSITE Ore crushing and stockpile installations alongside the railway loading facility at the original Goldsworthy mine.

plane everyone began to give the Aussie salute. Red earth, low mountains (Hamersley Range) with spectacular gorges. These mountains are virtually made of iron ore. Mount Newman mine, a partnership of Broken Hill, Amax and CSR, is one of the biggest iron deposits in the world, producing 33–35 million tons a year. The other big company here is Hamersley (CRA), whose Tom Price and Parabuda mines produce 35–38 million. Plane then took us to Port Hedland, where Newman and Goldsworthy ship their ore to Japan. Drove fifty miles inland to Goldsworthy itself. Huge, monsoon-like clouds built up, H-bomb fashion, into enormous sky, with occasional flashes of lightning. Does this every afternoon at this time of year. Flash set fire to the highly inflammable Spinifex grass (beautiful light-green colour) and tremendous plumes of smoke arose. Hint of cyclone, the great scourge here, especially from December to March. They normally pass over in about eight hours and don't cause much trouble, but in January 1980 one circled over

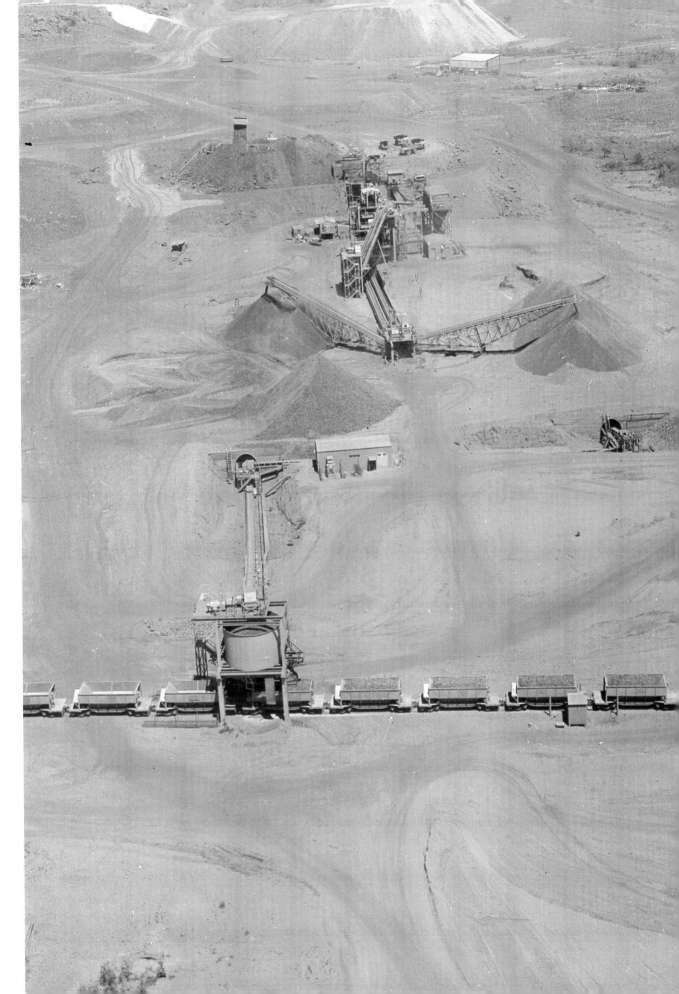

Iron ore from Mt Goldsworthy is exported from Port Hedland to customers in the Pacific basin, principally to the Japanese steel mills.

Goldsworthy and destroyed thirty per cent of the township. Almost all the houses were damaged. Soon after, another one hit Shay Gap. I found the temperature hot enough, mid thirty degrees, but usually it is much higher. In the Goldsworthy-Shay Gap area they claim they have now beaten the Australian record for a hundred consecutive days with temperatures over a hundred degrees fahrenheit.

At Goldsworthy met Alfred Kober and Derek Miller. Kober is General Manager in Perth but spends half his time up here. Came up here in 1965 before mining started and has virtually created the present operation. Miller is general operations manager. Goldsworthy has maximum capability of eight million tons but produces five million annually. Mount Goldsworthy itself, lease area A, has produced fifty-five million tons but is now mined out and closed since end of 1982. All the ore now comes from Shay Gap, area B and Sunrise Hill. But they still have their workshops here, plus railway operations and power-station, a total of 100–150 people, with 400 at Shay Gap, 250 in the Port Hedland area and 50 in Perth. The township at Goldsworthy, once 1,200 at height of iron boom, now has only 500 people, and falling – has air of ghost-town already. It costs a lot to run. The company does everything – repairs, drainage, sewage, water, power, education, entertainment. Rents are nominal. Ideally they would like to close it down but this would mean laying a power-line from Shay Gap to Port Hedland. It would cost millions and mining future is too uncertain to justify an investment on this scale. Went up the old iron-ore mountain, a grim and majestic sight. From the top of the empty mountain, 600 feet above sea-level, to the bottom of the pit is 1,100 feet, all mined in open terraces. It is gradually filling up to the top of the water-table, and they have cut up the surrounds to promote growth of vegetation. It already has a strange beauty and in time will be spectacular. The colours, especially the greens and reds, are immensely rich and varied.

From Goldsworthy drove along the line of rail, on good dirt road, about sixty miles to Shay Gap. It is in the middle of a glorious red rock formation – you can see the horizon of the iron orebody outcrops on the side of a line of low, flat hills. The new township is charming, in the middle of little red hills, with masses of trees and streams. It is cunningly designed so that the middle is free of cars and a mass of flowers, shrubs and trees. A tiny boy told us he had just seen a snake, 'that long' – stretching his arms. Guest house is admirably provided and we cooked ourselves steak dinner. Afterwards strolled around township in warm, enveloping tropical night with huge orange full moon and vast array of stars.

Up at 6 a.m. and tour of mine. Found No. 7 work-area by following the ore horizon. Art of open-cut mining is structuring the cut so that you even out the waste material, here mainly sandstone, you have to remove to get at the iron ore, so you don't 'dig yourself into a corner'. The other art is to avoid cave-ins of the walls. One side of No. 7 is firm and stable, so the angle of mining is more acute. On the other, sandstone side they go down in a series of concentric forty-foot terraces they call benches. When the sandstone bench-works crumble they tidy them up with front-end loaders. The loaders also take up the smaller ore-falls where mobility is required. For the big quantities brought down by blasting they use a cu.m electric shovel of which they have four. They are all old and two are out of action at any one time.

They have automatic drills to bore holes for blasting. Normally they blast at 11 a.m. or 3 p.m., but they blasted 7 a.m. this morning at No. 7 to get high-grade ore as they are short of it for their mix. The high-grade is known as a 'sweetener' and they mix it with a low-grade to maintain even average. They test quality of ore constantly. At another sweetener pit I watched the drill, a huge oblong machine with a fifty-foot gantry, and saw the men load explosives into the drill-holes. They put low-energy explosives, basically a fertiliser of ammonium nitrates, in each hole, plus high-energy detonator. By itself you can set a match to it and it won't explode, so it is very safe in the explosives truck. But with a detonator it goes off well in confined space.

Economics of the mine depend in great part on distance from ore-face to the primary crusher and railhead. They have found thirty to forty million tons more than they originally thought was here, but mining it has meant moving further and further away from the crusher. We went to pit at far end of range which is ten kilometres from crusher and at limit of profitability. The haulage is done by twenty giant Haulpaks carrying 108 tonnes of ore each. These monstrous trucks have limited visibility and so absolute right of way in all circumstances. Saw spot where, last August, a small truck didn't see a Haulpak's indicator and drove into its turn on the blind side. It was crushed flat. The Haulpak did not even notice – just had a small scratch on its fuel tank.

Some of the Haulpak drivers are women. The convener of the truck drivers is a woman. I was told the girls make better and more conscientious drivers than men, 'but we have to keep them out of danger situations as they tend to panic a bit'. They also made good train guards or 'observers' as they are called. As trade assistants or apprentices in electrical and engineering work they are less popular, as 'they can't do the heavy lifting which is a crucial part of the work'. One tough lady caused a three-week strike by socking a male colleague and breaking his jaw. But I was told: 'It is well worthwhile employing unattached women. It helps with the unmarried men, who drink less, etc. in consequence of a more active social life.' Some ore firms are plagued with strikes. Goldsworthy has the best strike record in the industry and one of the best in Australia – 'but we work on it'. They have an elaborate system of quarterly meetings and a representative body which runs many aspects of the township, a potent source of labour unrest. Brian Murray, who runs the port area, told me about twenty-five per cent of his time is spent on union negotiations. Miller says it is up to eighty-five per cent of his. But worth it – one tiny stoppage can have a cataclysmic effect on productivity. If three men stop unloading the train at Port Hedland, the knock-on effect backwards can stop the stockpile men, the primary crusher, the Haulpaks, right back to the mine-face.

Visited primary crusher on edge of escarpment. It is like an enormous steel pestle and mortar. At the moment they have 13.3 million tons of high-grade ore left at Shay Gap and Sunrise Hill. It will be mined out by June 1987. So they are thinking of building a beneficiation plant. At present they mine 1.6 to 1.8 tons of waste for one ton of ore. The plant would reduce waste to 0.75 tons. So they are building a pilot plant to see how efficiently it can operate. If it works well, they can mine a further 100 million tons-plus here, extending the life of the mine well beyond 1987, into the 1990s. By this time they will have opened up Area C. There, in the Marra Mamba formation, they have found iron-phosphorus ore of up to 2,550 million tons, with reserves of 245 million tons over sixty-two per cent. But it will need very costly developments – a 335 kilometre railway to Port Hedland, a township and expansion of the port area, costing $(A)2,000 million altogether. This would expand output to a minimum of 12.50 million tons a year. They want this on-stream for when China enters the market in a big way in the mid-1990s. So a beneficiation plant for Shay Gap ore would bridge the gap and keep the company operating.

From the primary crusher at Shay Gap the ore goes into rail-trucks. Samples are taken at a laboratory in a shed near the train-loader and results phoned through to Port Hedland an hour before the train arrives. The trains are made up of fifty-six trucks carrying seventy-five tons each, making 4,200 tons on each train; with five trains a day the throughput is 21,000. My diary continues:

Shown round port area. When train reaches here, ore has already been through primary crusher. The trucks open their bottoms and ore falls underground into a pit and is then taken by conveyor to a stacker, with swinging arm, which drops it on a long stockpile. Under the stockpile is a reclaiming tunnel with sixteen gates so they can select and mix ore from any part of the stockpile. Conveyor takes the ore into the crushing plant, where it is first screened and divided, then put through secondary and tertiary crushers.

The plant looks a mess because it was expanded in two million-ton jumps without interrupting production – like Topsy, it just growed. We thought in terms of four, five or six million tons as world beaters in those days. Actually, it is highly efficient, basically a simple operation but with a great deal of expensive hardware. Everything is covered in bright maroon-coloured dust. I talked to Gerry Sagaram, a chemist and Grade Control Engineer, who does the samples. They sample at every stage of the process, chemical samples every 250 tons and physical samples every 500 tons, and then adjust the process to get a constant product. NKK of Japan, which takes forty-five per cent, have their own inspectors on the spot and accept GML figures. The other customers pay ninety per cent on delivery and the remaining ten per cent when they are satisfied with the product.

Went to GML's loading terminal on Finucane Island. Opposite is the terminal for the much bigger Newman Mine. GML were first, so had to provide their own tug and dredge the channel. There is a high tidal range, over 7.8 metres, and channel is 15 kilometres long – being extended to 21 kilometres – so ships must leave on rising tide. The ships provide their own loading schedules, matching order, etc. to ensure trim, which are given to the GML operation people. They handle about fifty ships a year, and have had up to 110. The ships are getting bigger. The Newman company arrived in 1967 and dredged the channel to its present depth, and now, with two dredgers working, it can take up to 225,000 ton loads.

Newman envies Goldsworthy the placing of its loading terminal. An incoming ship can do a turn in the channel and be ready to take the place of a ship which has finished loading in forty minutes. At the Newman terminal, built to handle two ships at once, the first ship must be out of the harbour before another can enter it, and this means a big loss of time. Because of strikes, Newman has a low stockpile, and seven ships were anchored at sea outside, waiting their turn for the stockpile to be built up.

Goldsworthy Mining is a very cost-conscious and efficient operation, in a highly competitive industry. The managers admitted to me they are worried by the threat from Brazilian iron ore, whose enormous new Amazonian deposits are now linked to a new port on the coast by an 890-kilometre railway. But when I was there they were confidently awaiting the feasibility study on the proposed beneficiation plant, due at the end of 1985, and plans for opening up the Area C reserves in the Opthalmia Range, west of Mount Newman. Once the infrastructure is in, the operation should be good for a hundred years, and it is very probable that Gold Fields will still be producing iron ore in Western Australia in the twenty-second century.

It is, indeed, in Western Australia – and, to a lesser extent, in Queensland – that the real future of Australia lies; and both Gold Fields and RGC are well placed to share in it, for RGC has a commanding position in the mineral sands industry, both north and south of Perth. Outside the industry, it is rare to find anyone who knows much about mineral sands – rutile, zircon and ilmenite. The three are usually found together; occasionally you get ilmenite on its own or with a little of the other two; rutile and zircon occur together and generally with a little ilmenite. Ilmenite can be upgraded into a rutile substitute, and is then known as synthetic rutile.

Ilmenite is common: it is liable to be found wherever there is a coastline. The other two are more rare. In Australia all deposits were originally eroded from hard rocks inland, taken down the river systems onto the coast, and then redeposited between the tidal zones on beach strandlines. Mineral sands are heavy minerals, much heavier than sand. Tides perform a process of natural separation, and lenses accumulate; as the coastline recedes, old strata lines containing mineral sands deposits are to be found up to a hundred miles inland, especially in Western Australia. Alternatively, the minerals may be concentrated in dunes. So there are two kinds of deposits: lensal deposits, high grade, low volume; and dunal wind-blown deposits, low grade but large volume. The process of deposition continues: after a cyclone at Surfers' Paradise on the Queensland Gold Coast, you can see occasional dark patches on the beach. But people in the industry always try to avoid the word 'beach', for going to the beach is Australia's natural religion, and it is vital to the industry to separate its mining activities from beach-worship.

Early Australian prospectors noticed the dark patches on beaches, when really big south-east gales broke up fresh terraces of the black sand rock. They called them sniggers and mined them for gold. Auriferous sands were worked near the Richmond River in northern New South Wales and at Jerusalem Creek in Queensland during the 1870s; in the 1890s the sands were worked for lead. Then, in 1912, two American metallurgists used mineral sands to produce titanium dioxide as a rival to lead and zinc as a pigment for white paint. A further use was found in manufacturing welding-rods. The industry spread to Australia in the 1920s but it was not until the Second World War that it took off. Its creator was an ingenious Hungarian called Joe Pinter, who came to Australia in 1939, set up a welding-rod factory, found he had to manufacture his own raw material, and put together his own mineral sands separator using a car gearbox. Thereafter he built a series of processing machines, including electrostatic and electromagnetic separators, and in 1946 set up Associated Minerals. Fifteen years later, in 1961, Consolidated Gold Fields acquired a fifty per cent interest in Associated and expanded it into the biggest firm in the industry.

The industry is highly cyclical, following swings in the entire economy, especially housing-starts, for its main outlet is the paint industry. It boomed in the late 1950s, again in the 1960s, dropped sharply in the 1970s and is now doing well again. It has been a cut-throat business. Ned Wilson, at RGC's Perth headquarters, told me of legendary 'Border Wars' in the 1950s 'Some of Pinter's original leases were only ten chains wide, and there were arguments about rival mine access roads which had been built side-by-side. People fell or got pushed into dredge-ponds. There was a famous fight against the Murphy family in 1957. One of our D-8 Caterpillars got locked in combat with one of theirs – it was known as the Battle of the Bulldozers. I ran messages, being a lad, between the front and HQ. The Murphy's lost on that day – the bulldozers were damaged. Bill Bailey, our mining superintendent, was buried up to his knees standing in front of the bulldozers.'

The industry is calmer today, partly because it has to be highly organised to meet stiff competition from abroad, chiefly from Sierra Leone and South Africa. It also has to be on its best behaviour since it faces some of the strictest environment laws in the world, imposed by the Australian Federal government and the States of New South Wales and Queensland. Indeed, for all practical purposes RGC has now been driven from the east coast, where the industry originated, and its two production centres, Capel and Eneabba, are both in Western Australia.

Len Skelton, Executive Director of RGC, told me:

In the Capel area, south of Perth, the predominant mine is ilmenite. In Eneabba, north of Perth, you get a formidable array of all three minerals. The process of separation is all physical, rather than chemical. Zircon and rutile have different electrostatic and magnetic characteristics – so you can separate them by magnets or high tension electrostatic separators. The conventional mining approach is by dredging. You

Restoration is routine following the extraction of mineral sands by AMC in Australia.

create a pond. The aim is to dig to the lower limit of the ore body, while desludging the tailings behind you. You take out about six to ten per cent of the sand. Then there are physical separation processes to concentrate the material on the mining site. The mining process at Eneabba is dry – no dredger but bulldozers and scrapers. It is a more expensive operation than dredging but is used where ground conditions or other constraints prevent the use of a dredger.

The rutile product we sell contains ninety-five to ninety-six per cent titanium dioxide. It is processed further by our customers. The bulk of the rutile is used for pigment. The industry expanded in the 1950s when it was recognised lead paint was dangerous – kids ate the paint off their cots – and titanium pigment, by contrast, is non-toxic. It's also more opaque, so its got better covering capacity. As a flux in welding-rods it helps the steel to flow steadily and stops oxidation of the weld. Its more exotic use is as titanium metal – we boast about that but the quantity is quite small.

Rutile sells at $(A)500–600 a tonne. Zircon is a much lower value product, about $(A)160 a tonne and used mainly in ceramics, and the foundry industry as a foundry sand. Ilmenite is another source of titanium but cheaper, $(A)50–60 a tonne. It can be converted for pigment usage but you need huge quantities of sulphuric acid and this creates problems of waste disposal, so there are significant environmental problems. However, we have developed a commercially very successful process for synthetic rutile creation from ilmenite. This is our big hope for the future – we can turn our large reserves of ilmenite into a much more valuable material.

Synthetic rutile sells at $(A)300–350 a tonne. I myself would prefer the synthetic product – to run a pigment plant it is easier to use since, being manufactured, it is absolutely consistent. But people are used to natural rutile so they pay more for it. All the same, we are now making enormous inroads into the pigment market. We have two synthetic plants and are building a third. The reserves of natural rutile are being rapidly depleted. We used to supply from the east coast of Australia ninety-five per cent of the world's demands. Now the coast is exhausted or environmentally closed. It will be hard to come by in another twenty-five years and I think that by the decade AD 2,000–2010 people will be paying $(A)1,000 a tonne in today's terms!

RGC's mineral sands' people are enthusiasts. But because of its ups and downs, the industry did not have a high reputation at Gold Fields in London, at any rate until recently. Its efforts to contain and even reduce costs, while meeting exacting environmental requirements, are heroic. My diary records:

Rose 6 a.m. Drove with Wally Dawes, finance manager of AMC, from Perth to Bunbury along superb coast road to south. Then thirty kms. further along to Capel mine. Met Colin Brown, Scots manager – from Richard's Bay. He told me he uses his workforce two days a week on mining, five on processing, to avoid cost penalties on electricity – power is a significant part of their costs. They get the sand out by a bucket-wheel excavator, moved from one of their closed-down east coast operations – it is enormous, but small by comparison with the latest ones. The concentrator plant is moved periodically to be near the workings. It takes sands containing fifteen per cent heavy minerals and upgrades it in the gravity concentrator to ninety-two per cent – this is eighty per cent ilmenite, ten per cent zircon and small amounts of other minerals. These are then separated. They remove ilmenite, which is very magnetic, by cross-belt magnets. Then they separate the conductors and non-conductors – once you know the properties of the minerals it is a simple process needing perseverance and repetition.

The object of the synthetic rutile plant is to remove iron from the ilmenite and produce titanium-rich material. They were the first to produce synthetic rutile as a commercial product in 1969, in their small 10,000 tonnes a year Plant A. Then in 1974 they built their 30,000 tonnes Plant B. The Plant A has been mothballed and rehabilitated three times – is once more operational. Their strength is in their synthetic process; they utilise coal reductant and an accelerated corrosion process. Their competitors have to use a more expensive route. They have coal of good quality, close by. They have a lab with first-class chemists and metallurgists, all working on improving the process.

To meet environmental requirements they mine as follows. First they remove the topsoil and stockpile it. Then they dig out the sand in 100-metre strips using a bucket-wheel excavator. The material goes to the separator plant and the tailings into a dam, eventually covered up by the topsoil and used for farming land, though some excavated land is used as lakes for process-water. They fill the pit, put on a layer of topsoil, seed it, then put on cattle. They have a herd of 150 cattle, bought from, swopped with or sold to local farmers, with whom they are popular as they let them use their bulls free. The land was leased from the state forestry authorities and was previously unusable scrub. It costs $(A)2,000 an acre to rehabilitate – price of land locally is $(A) 1,000 an acre.

At the Eneabba mine, 190 miles to the north of Perth, rehabilitation is more difficult because of the extreme aridity; but it is carried out with great persistence and growing success. The truth is, as I discovered time and again on my travels, that once miners become persuaded of the case for rehabilitation, and are given reasonable funds, they become very keen indeed and extremely expert; they often know more than the environmentalists themselves, especially about local conditions. My diary goes on:

To Eneabba with Alex Kidd, Market Administration Manager. Went first to nurseries, where they cultivate more than 300 bush plant varieties. They take off the bush as well as the topsoil, stockpile it, and then after mining spread bush on area for rehabilitation because it contains seeds. If the plant-count turns out low, they top it up with seeds from the nursery. Lack of water slows the process – takes about seven years, depending on winter rain. Aerial photos taken 1981–85 shows the progress. There are some plants in flower at any one time, but the mass of colour is in their spring, September–October, when Eneabba mine is a blaze of yellow and orange. There is a lot of wild life. I saw emus, as big as ostriches, stalking all over the mine area. A few kangaroos.

At Eneabba the actual mining is done by a contractor with scrapers and front-end loaders. Everyone agrees that the mine has been transformed since the arrival, two-and-a-half years ago, of their new manager, Dick Scallan. It is now highly efficient, safe and clean. Went round separation plant – various spirals, magnetic separator, a new one from Germany (not working perfectly yet), another from the US and the most versatile one made here. Then next to drying shed – two heated rotary kilns to get the moisture out. Then zircon shaking air tables. The mined products come out to go by rail to Geraldton. It is a great grievance that the government will not allow truck transport. Sending by rail adds enormously to production costs. They have to get their water up from six hundred metres – that is another big part of costs. Lunched with Dick Scalan, on fresh crayfish he had just caught. He is a fervent Catholic and took me to his private chapel, where mass is said daily. He has a beautiful cockatoo called Joey, which he bought from a crane driver. It used to swear terribly, but his wife cured it. It squawked loudly at us: 'Praise the Lord! Praise the Lord!'

Scallan comes from South Africa. He told me 'When I came here I was very apprehensive after my South African experience and wasn't sure if I could handle Australian trade unions. But management is management all over the world and the same skills apply. I was dismayed by the high turnover of men, seventy-five per cent a year. Now we have got it down to twenty-five per cent. Of course, many can't bear it here. You must remember that with a thirty-eight-hour week, leisure occupies more time than work – so what people can do with their leisure is in some ways more important than conditions at their workplace.' He is very keen to hire family men, especially if they have large families. He is getting in Poles, for instance.

RGC's mineral sands' division is very confident at the moment. But its boss Peter Cassidy did not conceal the competitive rigours when he talked to me in Perth:

At one time Australia had eighty to eighty-five per cent of the market. It's now declining. In zircon, over the last five years, Australia has gone from seventy to seventy-two per cent of the world market to about sixty per cent – this is because of the big Richard's Bay operation. For rutile and ilmenite we once had over ninety per cent – now it's fifty per cent. The competition comes from Sierra Leone, Richard's Bay and our own operation in Florida. Richard's Bay will always sell zircon at $(A)10 a ton lower than Australia. For zircon there is a good balance between supply and demand. For ilmenite we are no longer a major producer in our own right – it goes to synthetic rutile mainly.

There is no problem in selling rutile at the moment, but when demand is poor competition comes from all sides. Richard's Bay coming on-stream with such huge quantities of material has had the biggest single impact on world markets and competition. The industry was a sick little

Mining mineral sands at Eneabba, 150 miles north of Perth.

business in the early 1980s. In 1984 there was a big improvement, 1985 is excellent and 1986 will be good, too. This is the upturn in the world economy. The paint pigments industry is very strong and the metal market picking up. But welding, tied to heavy engineering such as shipbuilding, is still slow. Pigment is now at capacity – depends on housing, cars, paper and, to a lesser extent, whitening. Metal industry is related mainly to America's defence programme and the aircraft industry. Reagan's decision to revive the B–1 Bomber had a major impact on demand for titanium metal. Each B–1 consumes ninety tons of titanium ingot. Titanium sells at $(US)4.00 a pound. A pound of pigment sells at only seventy cents a pound.

About seventy per cent of our production goes into pigment, ten per cent into metal. When the upsurge started, we had surplus stocks. We reverted to full capacity and sold off stocks. Unlike other divisions of RGC, which sell on published prices, we operate on marketing demand and negotiate prices. We are halfway between mining and manufacturing – a lot depends on marketing. In the case of zircon we produce twenty-five per cent of world production. In the case of rutile and synthetic rutile we produce twenty to twenty-five per cent of total world production. So we have a major position but not big enough to dictate prices. Ours is a very complex industry, with a very wide variety of end-uses and a wide variety of minerals which go into those uses.

We once had a very poor relationship with RGC. But technical changes have reduced operating costs and we are now able to take full

Photo micrographs of zircon (above), rutile (centre) and ilmenite (below), the principal products of the RGC mineral sands division.

advantage of better times. We don't have much feedback from London – it's filtered through Sydney – but 1985 should produce the best profit figures in our history. Over the past three years a lot of time and effort have been put into improving operations. The rationalisation which has taken place in the Australian industry as a whole has also helped. There used to be a large number of companies. Now there are only a handful, each associated with big mining organisations. So you now have highly professional managers who take a long-term view and do less hip-shooting in reaction to steps taken by rivals.

Much of the division's future hopes lie in Florida, where it is mining sands in the middle of a pine forest at Jacksonville, twenty miles from the present shoreline and 100 feet above sea-level. RGC moved there in 1980 and started to operate profitably for the first time in January 1985. Len Skelton told me:

The scale of our Florida operation will be very big. It could become a big synthetic rutile operation, as there is plenty of ilmenite. We could possibly treble our reserves in the Florida region. We are building a third synthetic plant in Western Australia because we're not absolutely sure of adequate Florida reserves, but the fourth plant could be in the US – and it could be a big one.

Michael Shepherd, the top mineral sands geologist in Perth, told me that their exploration in Florida is an Australian effort, though they have two people seconded from Gold Fields Mining Corporation exploration headquarters in Colorado, where they look upon mineral sands as a peculiar Australian expertise. He told me:

There are four good reasons for going into Florida. Firstly, they are less hostile environmentally. Secondly, there is political stability. Thirdly, it is geologically promising. The industry started in the 1930s as in Australia but deposits were not so rich so they were slow at getting going. Fourthly, inexpert mineral sands mining in the US gives opportunities for Australian skill. In the US, the Crown land situation we suffer from here does not exist – we are dealing largely with private land. But there are problems in Florida owing to the bad behaviour of the phosphate industry, increasing the radiation content in the soil. Mineral sands operations take out only five to six per cent of the contents, so the disturbance is minimal and restoration can be pretty complete. Phosphates take out a great deal, and are pulling out of Florida for North Carolina. But in consequence the Florida authorities have become very conscious of environmental factors, especially water-use, and I fear they may over-react.

In any case, he added, the aim of the exploration programme was to look at the entire East Coast of the United States, from New Jersey southwards.

Mineral sand mining is a quintessentially Australian industry. It is very much a world on its own, with its peculiar jokes and jargon. In techniques it is the world leader. But it is a disturbing portent that RGC's mineral sands

division, the most go-ahead firm in the industry, should have virtually ceased exploration for new deposits in Australia itself, and should see its long-term future as mainly abroad. The story of how RGC was inhibited in its mineral sands activities in Eastern Australia, by both State and Federal government, and finally decided it was prudent to pull out altogether, is a cautionary one. There is a clear danger that Australia, in pursuit of an ultra-clean environment and a leisured way of life of which golden beaches are the symbol, may undermine the economic foundations on which its cherished prosperity rests. At the State level, one or two wiser spirits perceive this danger. It is significant that the New South Wales government, which has succumbed most easily to pressure from the environment lobby in the past, has set up a task-force to devise means of luring the mineral sands industry back into the State. They will not find it easy.

Moreover, though State government may be uneasily aware of the need to nurture industry, the Federal government has been much more ruthless in its pursuit of a pollution-free utopia, using or abusing Federal law to overrule the States. Thus RGC was prevented from mining the sands at Moreton Island in Queensland, despite the willingness of the State authorities, by the refusal of the Federal government to grant export licences. Again, the Federal government overruled the Tasmanian State decision to build the Gordon River dam on the grounds that, as the area was

The environment need not suffer as a result of extracting Australia's mineral wealth. Here, an Associated Mineral's dredging plant in the 1960s is recovering rutile and zircon from sands along the beach at Cudgen South, New South Wales. The floating dredge and its pond move along the beach as dredging proceeds, bulldozers then restore the worked-out site to its natural state. AMC discontinued all its east-coast operations in the early 1980s.

classified as 'wilderness' and registered with Unesco, the matter was a foreign policy issue and the responsibility of Canberra. The Federal government is also raising increasing obstacles to the expansion, or even the survival, of the mining industry by developing the legal concept of Aboriginal tribal lands, and by devising a mass of restrictive legislation in consequence. This is already a severe disincentive to exploration and development in the Northern Territories, and a threat in Western Australia, where the mining future of the country chiefly lies.

Meanwhile, costs not only in mining but in all industry are being raised by ever-increasing statutory requirements for the protection of employees. For instance, at Capel, which employs no more than 130 people, I found management were now obliged by law to employ a full-time 'Senior Occupational Health Engineer'. Enforcement of new noise-control regulations, effective from October 1985, adds further pressure on prices. Companies are now liable for accident claims from employees injured travelling to and from work. They face the further threat of claims under the dubious heading of 'repetition strain injuries', which at present applies only to clerical workers but will ultimately be extended to all occupations. In its craving for the perfect life, Australia is inclined to assign statutory responsibilities for all the ills whereof the flesh is heir to, and to send the bill for damages to private industry.

No one who travels in Australia can doubt its enormous economic potential. As a frequent visitor, I am convinced that the exploitation of its mineral wealth, for instance, has only just begun. The opportunities are there in great abundance. But government must now be reckoned the most formidable single problem which the entrepreneur faces there, not least when it reflects the suspicious and even hostile attitude to foreign investment which has become an Australian characteristic. With the re-organisation of its Australian interests at the beginning of this decade, Gold Fields has gone a long way towards meeting all reasonable wishes that the country should control its own natural resources. It is eager to anchor itself into the local culture. Not long ago, for instance, it commissioned the distinguished Australian painter, Sidney Nolan, to paint a series of landscapes of RGC properties. He visited them all. At one mine the manager arranged the date of the visit, but at the last minute there was a sudden crisis and he was called away. He forgot to tell the Assistant Manager. The latter was rung, on the day, by the receptionist: 'There's a bloke here called Sid Nolan. Says he's a painter.' The Assistant Manager, a busy man, replied: 'Use your head, girl. Send him to the paint shop.'

Michael Lynch, a Gold Fields' executive then responsibile for liaison with Australia, gave me his thoughts on RGC:

In our iron-ore business we are looking fifteen years ahead. We are trying to position ourselves in relation to other huge operations in the Australian–Far East area. Over the kind of time-scale we deal with it is vital not to rule out future options. The future, as we have found out from bitter experience in the past, is so difficult to forecast that at some time you have to commit big acts of judgement. The information is never

OPPOSITE RGC has a 60 per cent interest in Pine Creek in Northern Territory which came to production in 1985 and produces about 2 tonnes of gold a year.

RGC has its headquarters in Gold Fields House, Sydney Cove.

adequate – at the end of the day someone has to make a qualitative judgement. So area managers have to be given a lot of rope to exercise it. You hsve to have a virile management and to do that you have to give them their freedom.

Costs have escalated for various reasons – worse deposits, chiefly – to the point when new iron and copper mines cost one or two billions to build. Hence in the present climate we have needed to switch to smaller projects where the risk is smaller – except in South Africa, where they are used to enormous mines built at huge cost. It is a truism that present managements survive on past management's successes. Your immediate control over what is happening today – which is what the newspapers concentrate on – is very small. There is not much you can do with a mine operating now. Exploration is an imprecise science and lots of luck and hunches are involved. A geologist's working life is designed to find one mine – if he is lucky. But Australians have a tremendous investor interest in exploration. They love to put their money into mining shares. The country is heavily dependent on exports and on capital inflows. It suffers from drought. Unemployment levels are high. There is a lot of trouble from government – the rejection of uranium, Aboriginal land-rights and so on – and many companies are discouraged by the investment rules. However, we have now got the structure of the Australian company right. It's been costly and painful. But if the exploration is OK, everything will come right. We have to take a long view and not until the 1990s will any results appear which will influence Gold Fields at the centre. At the moment they get lost in the noise.

In Sydney, in RGC's spectacular skyscraper overlooking the harbour, Max Roberts told me he thinks the relationship with London is basically sound. 'There had to be devolution', he said to me. 'If not, it wouldn't work. If I'm not efficient, fire me. If I am, then leave me to do things. It's not London's job to act as policeman. But given that, there ought to be free and open communication between Sydney and London, and that we have.' Roberts has now handed the duties of Chief Executive to Campbell Anderson, who has the prime task of achieving RGC's full potential. Anderson, like Roberts, came to RGC from the oil industry, with executive service for Burmah Oil in the United States and Britain. As for the future of RGC, Roberts summed it up for me as follows: 'We are strong in tin. We are getting out of copper. We are happy with mineral sands. We are going into gold. We are getting a piece of energy. By rights, we should have been a big company by now. Many mistakes were made in the past. But we are now on the right road.'

ABOVE The Craig Yr Hesg
quarry near Pontypridd,
Glamorgan, in the 1920s.

RIGHT The ARC Chipping
Sodbury quarry has been
worked for over a century.
This photograph taken in the
late 1930s shows R. W. Amey
second from the left.

6 ARC Limited:
A British Success Story

THE most far-sighted and successful move by Gold Fields, since the valiant decision to develop the West Wits Line taken in the early 1930s, was the campaign to make the company a major force in the British construction materials industry. It is a campaign which, in less than twenty years, has established Amey Roadstone Corporation (officially known as ARC Ltd since 1 April 1986) as one of the soundest and most enterprising businesses in Britain and from 1984, when its profits exceeded £55 million, the biggest single contributor to the prosperity of the entire group. It is a remarkable British success story, and one of which CGF is rightly proud.

The story goes back to the mid-1960s, when the profits from the West Wits mines were rolling in and Gold Fields, as part of its plan to diversify from gold and South Africa, was looking for long-term investment opportunities. The directors took the view that CGF ought to have a major British interest. Over the years it had held shares in many British companies, but it had never developed a self-sustaining centre for growth. That was now the object. But what should it be? Britain was no longer a force in metal mining. Coal was nationalised. North Sea oil and gas was not yet a prospect. The choice fell on aggregates. Superficially, gold-mining and quarrying appear to have much in common. Actually, they have very little. In gold-mining there is no problem in marketing the product: the art lies in finding and recovering it. Quarrying, by contrast, requires relatively simple physical skills but increasingly sophisticated marketing ones.

What the two industries do share, however, is the long-term requirement for large reserves. It was this common factor which led Gold Fields to perceive its opportunity. The essence of modern business planning is to detect changes created by the interaction of economic forces and government decisions before they become overwhelmingly evident. During the post-war period, the south of England established itself as the centre of natural economic growth in the United Kingdom. That in itself put pressure on local aggregate resources. The extension of planning permission to aggregate extraction increased the pressure. By the mid-1960s, the building industry was aware that reserves of exploitable aggregates would be increasingly hard to get, and the big firms, such as Tarmac, Redland and Ready Mixed Concrete, were securing them by buying up the small family companies which chiefly possessed them. These little firms suddenly began to realise they were worth a fortune. Gold Fields decided to move in just in time, before the price of reserves became prohibitive or the supply ran out.

The official name of the company was changed to ARC Limited, which reflects common usage and recognises the continuing expansion in the scope of the business.

RIGHT The ARC head office built in 1980 at Chipping Sodbury, near Bristol.

BELOW Premix in the early 1960s. Now, ARC Premix is the second largest supplier of ready mixed concrete in the country.

BELOW Aggregate production remains at the heart of ARC's business. Planned reserves are sufficient for over 100 years.

The purchasing programme was entrusted by Donald McCall to Gerald Mortimer, who had just returned from service as the first General Manager of Goldsworthy Mining. Mortimer was to be the first Chairman and Chief Executive of the new ARC, a position later filled by Rudolph Agnew. The first big purchase was Greenwoods at St Ives in Cambridgeshire which had sand and gravel operations chiefly in East Anglia. The following year Gold Fields acquired control of Amalgamated Roadstone Corporation, which became the core of the new division. In 1972, after a struggle, it secured control of the Amey Group. Thereafter the new firm, Amey Roadstone Corporation, was big enough to develop its own resources and expand. The first few years of the new investment were not auspicious. The CGF board grew very restless. The industry was riddled with price collusion. In the second half of the 1970s it contracted by thirty per cent, with the sudden fall-off in the building of town halls, hospitals, schools and other publicly financed projects. By 1979 the industry had shrunk to two-thirds of its size. Yet the effect of this rapid concentration was to raise profits. Six big companies emerged (plus many small ones). ARC's profits soared from around £1 million in the early 1970s to top the £50-million mark a decade later. Moreover, the growth of ARC fitted in perfectly with Gold Fields' decision to expand its aim and philosophy from mining to the provision and beneficiation of natural resources. After North Sea oil, aggregates form the largest natural resources industry in Britain still in private hands – and Gold Fields has a key portion of it.

Moreover, ARC is a peculiarly satisfying company with which to be associated. It is undoubtedly big business; it is modern-minded, it is operating at the frontiers of its own technology. Yet it retains much of the flavour and all the virtues of its origins in small-scale local enterprise. It is very British. Indeed it has a distinctive West Country accent. Its headquarters are not in London but in Chipping Sodbury, near the site of a quarry it has worked for over a century, where in fact fine quality stone has been extracted since Roman times. It breeds a particular stamp of unit manager: robust, expert, down-to-earth, with strong local roots.

ARC was really born of the realisation that the age of rail was over and the age of road had come – or come again. In the eighteenth century Britain had pioneered the first advances in road construction since the Roman empire, but throughout the nineteenth century rail had carried all before it, and little was done to improve road-making techniques or to rationalise the industry which supplied it with its materials. The man who first saw the need for change was Sir Henry Maybury, an engineer who had started in railways, become an expert in road construction, and ended his public career as Director-General of Roads at the Ministry of Transport. When he retired, in 1928, he immediately set about organising what he felt was needed: a rationalisation of the road materials industry. He knocked together eight quarrying companies into one big one, the British Quarrying Company, the first major unit of its kind. The eight included John Arnold Limited, which ran the Chipping Sodbury quarry, the new headquarters; all the rest had histories going back to between 1872 and 1896.

Maybury's company was so successful that it started a trend. In 1934 a rock-asphalt specialist turned financier called Gatty Saunt knocked together four big quarries to form another major group, Amalgamated Roadstone Corporation. ARC was based on Cornish quarries and had a sea-borne trade. BQC was mainly in Gloucester, Shropshire and the Welsh marches, an inland outfit. Gatty Saunt realised it made sense to bring the two together and he arranged a merger in 1946. The joint company (ARC) flourished, moved into cement and cement-products, acquired additional quarries and reserves, and in 1964 made a profit of over £1.5 million. In 1967 it merged with another West Country-based company, Roads Reconstruction, but the new arrangement did not function smoothly and within a year the enlarged ARC was ripe to be taken over itself. With ample reserves in the south and west it was exactly what CGF was looking for and the deal went through, Gerald Mortimer taking over as chairman for the next seven years.

But CGF was not happy with internal growth or profits. In 1972 it bid for the Amey Group, a fiercely independent firm of similar size to ARC. It had been created by the Amey family, who still ran it; but it had gone public in 1960 and the Ameys now owned only 20 per cent of the equity. Amey was an excellent firm with a good profit record. It operated mainly in central-southern England, and had a substantial dowry of quarries and reserves as well as much expertise in the building materials trade. CGF had thought of taking it over as far back as 1967, but the Ameys had refused a deal. They had been offered other bids and mergers since but their terms had always proved too stiff. On 20 June 1972, CGF made an offer for the shares in a letter to shareholders, giving 11 July as a closing acceptance date. The offer was rejected by the Amey board, and the Chairman, R.W.Amey, in a letter to shareholders, said it was unwelcome. The battle that followed was short, hotly contested and very exciting. CGF promptly returned with an improved offer, for acceptance by 9 August. It valued Amey shares at 285p (they had been 174p the day before the first offer was known) and put

Brigadier-General Sir Henry Maybury, (1863–1943) founder of British Quarrying Company.

W.H.Gatty Saunt (1888–1965) formed Amalgamated Roadstone Corporation in 1934 and merged it with the British Quarrying company in 1946.

Amey's earnings at a price-earnings ratio of 27.8. The Amey board rejected this offer too and advised shareholders to do the same, arguing in a letter to them that the price-earnings ratio had been 33 when CGF bought ARC four years before. They further pointed out that profits for the year to end-December 1972 would be significantly higher, at about £4.5 million, and that reserves of aggregates were over 430 million tons.

The offer was tempting but the Amey board's arguments also persuasive, and shareholders were evidently very divided. But on the last day of the bid and almost on the last hour, CGF got sufficient acceptances to pass the fifty per cent mark, and by the time the offer ended had obtained 51.08 per cent of the ordinary capital. Ron Amey then graciously acknowledged defeat, accepted the bid himself and advised shareholders to do likewise.

Hence by the end of 1972 the nucleus of the present firm had been created, and thereafter rapid internal growth was possible, though ARC has also continued to expand by purchase. The components of ARC were essentially created by rugged individualists, many of them self-made entrepreneurs of humble origin. The three Arnold brothers who ran Chipping Sodbury branched out into quarrying from a Somerset cider orchard. William Amey was a small-scale market gardener, who moved into the business by digging his own gravel and burning his own lime. His son Ron was a local autocrat, who insisted that all directors clock into their offices Saturday and Monday, attend board meetings every Friday, and live within ten miles of Wooton (Oxfordshire), the company's heart. Douglas Cleaver, the man who made the ARC-BQC merger work, was another

William Amey, celebrating his 90th birthday, with his son Ron Amey and Rudolph Agnew (seated right), on 6 December 1972.

local countryman, who attended board meetings with two large dogs and insisted that board lunches should be built around a proper joint, which he carved himself. The vast majority of quarry managers in all the companies were, and indeed still are, local men who have come up through the ranks.

Given the high quality and the spirit of independence among the management of the properties CGF had taken over in the 1960s and 1970s, there was really no alternative but to pursue a policy of devolution. In fact ARC was well adapted to fit into Rudolph Agnew's evolving policy of transforming Gold Fields from an empire into a commonwealth. He himself told me: 'ARC is a very good example of the management philosophy I support.' But there is one important qualification. Unlike the associated companies in South Africa and Australia, and unlike Newmont, ARC is a hundred per cent owned by Gold Fields. Antony Hichens accepts that this is a 'contradiction'. He explains it as follows. 'Tax is one factor. We can offset advanced corporation tax on worldwide dividends against any British corporation tax, but not against foreign tax. The point is complicated to explain, but the financial advantages are compelling. Secondly, there is no great difficulty in getting good British management to run ARC. It's not like the GFSA people feeling they were working for a foreign – that is, British – holding company. There is a big psychological gulf between foreign and local ownership, and in foreign countries it is better to be seen to be locally owned. So this argument did not apply to ARC. Obviously, Gold Fields has a slightly bigger say in Roadstone and a more easily exercised veto. But there is a high degree of autonomy.'

A Roman pavement at the Stanwick quarry in Northants, excavated with ARC's support by the English Heritage trust.

Agnew believes that a greater degree of formal, structural independence for ARC would be an advantage. He told me: 'We have an extremely valuable asset there, managed by good, solid Englishmen, mainly West Countrymen – they don't speak the American business jargon. But as a hundred per cent owned company it doesn't always have the edge its competitors have in the growth area.'

All the same, ARC has been growing fast. In addition to the internal growth in traditional areas and the development of new products the company is still seeking substantial acquisitions. Blue Circle Aggregates added to the base business, Westminster Gravels also doubled the marine dredging, hastening the move to being the biggest factor in this rapidly growing business. The most recent acquisition, Bath and Portland Group, in addition to major stone reserves and production, gave a valuable impetus to the emerging property development business in ARC. This is a logical but difficult step in the progress of a company with large land holdings and experience in dealing with planning authorities. It is more soundly based than the American operation in that it already has a massive stake in aggregate reserves, which are in limited supply, and can now expand into processing and manufacturing, where the opportunities, at least in theory, are limitless. Humphrey Wood, who became Chairman in 1979, and Charles Spence, ARC's Chief Executive developed a strategy of expanding ARC further into the private sector. (Humphrey Wood stepped down as Chairman on 1 July 1986 and was succeeded by Charles Spence who now combines both roles.) Charles Spence explained to me his philosophy of growth:

> We are strong in the south of England – the concentration diminishes as you move north of the Midlands. In the south our opportunities to expand are restricted because of our already large market share. To grow without running into monopoly problems, we have to move north. But the population is moving south. So? We will take up a strategic position in the north if it offers. It is our intention to get more into construction materials which relate to the private sector – it is essential to do this. The object of the company is substantially to increase its size over three to five years. It's possible for ARC to double both here and in the US. You can only do that in the UK by getting into new products – construction materials over a broader base – and in the US by getting more into aggregates.
>
> In this industry there is no optimum size. But it's best to be very small or really big. We have been big for some time. The essential thing is to have the right structure for the size of the company you've got. ARC has a turnover of nearly £600 million a year. CGF has a capitalisation of £1 billion. You recognise the size of the company and build up a structure appropriate for that. The degree of delegated authority follows the same principle. Agnew allows a hell of a lot of autonomy – he allows the company to run in a way he considers appropriate. Within ARC, I have a number of directors responsible for functional aspects – for instance, a director responsible for aggregates. He in turn has four regions, divided into eleven areas. There is a managing director for each region, and a manager for each area. Then I have a main board director for pre cast concrete, for construction, for the United States, a financial director and an administrative director.

ARC enjoys a high degree of internal devolution. There is no other way to run the company. It was built up from a number of independent units, and they remain units – pits and quarries scattered around the countryside, often with manufacturing plants attached or nearby. The cost of transport is high in relation to the value of the product, so market areas are restricted. This is an intensely localised industry. It is also low profile – literally so, in a sense, for the industry is ruefully aware its workings may be messy and ugly and are unpopular, and every effort is made now to conceal them from the public. The ideal quarry now is one you do not realise is there until you enter it; the art of 'contouring' has evolved to make the business of digging up aggregates as invisible as possible.

It is low-profile in a wider sense too, for the construction materials industry is little known outside its own work places and board rooms. We take its services for granted. In a modern Western society, we assume without thinking that sewers will function, that rainwater will drain away, that rubbish will be taken and disposed of, that roads will be laid and mended, that airports and seaports will be available as required. Public attention is often sharply focused on the cost of these things. But the actual provision of the enormous quantities of materials needed to build and

maintain the basic physical infrastructure of a civilised, industrial society is rarely considered at all. The problems which confront the industry, the way it solves them, its changing technology, its triumphs and achievements – none of these things make the headlines but are quietly cherished in professional journals and house magazines.

Examining the industry, then, is to go behind the scenes of the affluent society. You see bits of structural machinery not normally exposed to the public gaze. Some of it is on a tremendous scale. ARC's biggest quarry, at Whatley near Frome in Somerset, has recently been doubled in size, to four million tons of carboniferous limestones a year and there are plans to double it again. The manager, David Yelland, runs a workforce of a hundred, but the site is enormous and you do not see many around – most of them work on maintenance or running the railhead. The industry is increasingly capital-intensive and the investment lies in the equipment. Every day, Whatley blasts down 10,000 tons of rock. It is then grabbed by a wheeled loading shovel of formidable size. This alarming custom-built monster weighs nearly a hundred tons and costs half a million pounds – its tyres alone cost £10,000 each, stand nine feet high and weigh over two-and-a-half tons. It uses twenty-five gallons of fuel an hour but it works a sixteen-hour day and shifts two million tons of rock in a year. It shovels the rock, seventeen tons at a time, into dumper-trucks which take it to be crushed. Whatley has three types of crusher: a primary, which is a vast, gyrating

ABOVE A blast at the ARC Chipping Sodbury quarry.

OPPOSITE Whatley near Frome is the biggest ARC quarry. It supplies south-east England by road and rail as well as serving local markets. The current redevelopment programme, the first stage of which was completed in 1986, will eventually enable output to reach 10 million tonnes a year.

LEFT The face at Whatley.

OPPOSITE The ARC sand and gravel pit at Somerford Keynes in Gloucestershire opened in 1964. It covers 380 acres and following extraction restoration is an integral part of the operation.

pestle and mortar; an impact crusher, which drops the stone on a fast-rotating mass of metal; and a jaw-crusher. All this is being replaced by a £12 million new crushing plant which by the end of 1986 will have four times the present capacity. The crushing process reduces the fragments of rock to ten different sizes, which are then available for making concrete or, in various proportions, for making coated stone or sixty or seventy different types of material. A lot of it goes out as sub-base for roads or drainage; occasionally for walling stone. British Rail transports 7,000 tons a day; 3,000 tons goes off by road. As the investment programme is completed, Whatley will

Three hundred years of Portland stone – St Pauls and a new British Telecom office. ARC acquired the Bath and Portland group, which included the Portland stone quarries, in 1985.

emerge as one of the biggest-capacity quarries in Europe.

All quarries are different. To work some, requires complex engineering. It may be necessary to tunnel under roads. Opening up the side of a hill can be very skilful work. Quarries vary greatly in size too. At Somerford Keynes, Martin Ellis manages an ARC sand-and-gravel pit which covers 380 acres and employs six men. Ellis told me:

ARC craftsmen training future stonemasons.

> We opened in 1964. All local labour in those days. We produce pea shingle (6 mm) and 10 and 20 mm, 1 inch reject, one grade of sand and plaster-sand. An excavator on a tracked drag-line gets it out and loads it into a CAT dumper-truck. They bring it up and feed the hopper. It goes onto a feed-belt and this puts it onto a conveyor, which takes it to a scrubber-mill where it's washed with water at twelve revs to the minute. Then it goes onto rotary screens, inner and outer, which divides the sharp sand from the gravel or shingle. That goes on a carrier which in turn goes onto a Niagara shaker-screen, rotating and pushing the material over different sized screens, shaking and sorting it into appropriate bins. The sand is pumped along pipes into the sand-tower, then onto lorries for delivery. The shingles go into a storage bin and are then fed onto lorries by conveyor.
>
> The demand for sand may lead to more shingle than we can sell, so then we crush the shingle. But I find that shingle always goes in the end. Digging can be a very delicate process, so as not to ruin your material. There's a lot of rule-of-thumb and human ingenuity – on-the-spot

RIGHT Transporting sand and gravel by conveyor from the working face to the processing plant at Lydd in Kent.

BELOW Processing plant and stockpiles at the Coln gravel pit near Lechlade, Oxfordshire.

technology. You have to develop instinctive skill in digging, which you get from long experience. Digging dry, you can see what you're doing.

I started in lorries. I can operate every piece of machinery in this quarry, and I sometimes have to, though it's against the rules. The men only respect a manager who can do their jobs at least as well as they can. You can't drive them. Not Gloucester lads, that is. You might with Berkshire lads or even Worcester lads, but not with Gloucester lads, you wouldn't. Since 1964 there's much more technical grading of material, introduction of British Standards and so on. A holiday shovel driver, who doesn't know where the bottom is, lets in foreign bodies, so your load is rejected. A customer will kick if he finds one stone in it. Quality control – that's the main problem. Coln Gravel at Lechlade has the answer – everything stored on concrete so you pick up only what you pay for. It always pays to use the materials at hand – lay concrete, lay good tarmac approach roads. It quickly saves you money. Often you can use your own men to do it.

These days, a quarry manager has to know as much about constructing amenities as extracting – water-skiing, nature reserves, caravan parks. I like to develop these as I go, it costs less to shape than going back in three years' time.

He deals with many bodies and authorities including, for instance, the Thames Conservancy, because he sometimes has to close a little tributary for a time. He has already restored much of the land he had processed, laying down fourteen inches of topsoil, then putting sheep to graze. He showed me his lakes, especially the fifty-acre one he calls the Big Lake. It has a huge population of coot and is much used by Canadian geese. 'And swans – I've seen up to a hundred swans on this lake.'

Many aggregate reserves, especially in the south, have already been exhausted or locked up permanently by development or made inaccessible by planning restrictions. Hence the search for reserves has moved out to sea, and ARC Marine, one of the fast-growing divisions within the group, has now become the largest supplier of marine aggregates to British and Continental ports. Operating from Southampton, it runs a fleet of thirteen sea-going dredgers which can carry 30,000 tons. This is another capital-intensive and highly profitable operation conducted by a tiny workforce – ten men per ship as a rule, and they do the discharging too. Once the ship has reached its dredging area, which is from twenty to a hundred miles from port but occasionally much further – the coast of East Anglia is one hunting-ground – it drops its drag-head or pipe, as they call it, about a hundred feet and uses centrifugal pumps to suck up the aggregate. The master controls it all from the bridge and as one of them explained to me there is a considerable art in keeping the ship steady in position during the three-hour suction process: 'The pipe must trail at the right angle while the ship steams slowly ahead. The great thing is, mind, not to sit on your pipe and break it.' A big ship like ARC Marine's *Deepstone*, which steams at thirteen knots, carries 7,500 tons and can discharge its cargo at the rate of 2,000 an hour, has an enormous annual turnover. Offshore aggregate dredging is, of course,

Restoration goes hand in hand with production at ARC sand and gravel pits.

The Arco Severn off-loading marine aggregates at an ARC wharf.

strictly regulated and breaches of the rules are sharply punished. But an efficient, large-scale operation like this can be very profitable. The main markets are naturally on or near the coast. A typical major ARC Marine contract was the delivery of 150,000 tons of material for land reclamation during the building of the new Townsend-Thoresen terminal at Portsmouth. But it is indicative of the growing competitiveness of marine aggregates that ARC Marine is now winning contracts to provide sand and gravel as far inland as the London M25 orbital motorway. It is also penetrating Continental markets, which provide a third of its business, and its Continental subsidiaries are expanding steadily. ARC is now the Number One in marine gravel and is building one new ship at a cost of £7 million and has another planned. It is clearly a natural area for growth. But the trade is highly competitive and good results are only achieved by constant attention to costs, great skill and *esprit de corps*, and heavy investment.

The biggest impact on the costs of quarrying – or rather on its profits from by-products – has been the recognition that the end result of quarrying, a vast hole in the ground, is itself a valuable and exploitable natural resource. It is an axiom of Rubbish Theory, now an expanding academic discipline, that the poor remain poor by throwing away their rubbish while the rich become richer by re-processing it. It could be argued that there is no such thing as absolute rubbish: some use can always be found for it. The great discovery of the modern aggregate industry is the marriage of holes in the ground to rubbish. ARC has thirty-five quarries and eighty sand and gravel pits, which annually remove over twenty million tons of aggregate from the ground. The resulting cavities are ugly and dangerous; they have to be rendered safe and if possible put to use. Some can be turned into lakes for fishing, boating and wildfowl. But the best solution is obviously to return them to agricultural usage if possible. That is where rubbish becomes valuable. British households alone discharge into the local government disposal system about twenty-five million tons of rubbish a year. Much of this rubbish becomes infilling for aggregate cavities and so the basis for fresh fields and pastures new. What is more, the local authorities will gladly pay the aggregate industry for taking its detritus.

The business of handling waste for landfilling has now become highly scientific. The object is to reduce in time the cycle which begins with quarrying agricultural land and ends when the area is restored to agricultural use – if possible richer than before. ARC gobbles up 480 acres a year; but it is actually putting something like 750 acres back into the system as old quarries are infilled and the depredations of the past redeemed. At some quarries and pits – ARC's Offham in Kent is an example – the whole cycle can be taken in at a glance: agricultural land awaiting quarrying, the quarrying operation itself, the dumping and covering of rubbish-landfill, and the restored land once more sown with crops or used for grazing. ARC now not only offers sites for final disposal but provides a range of services for getting there, including special vehicles and containers for both road and rail.

At Sutton Courtenay in Oxfordshire, for instance, it runs a huge operation in conjunction with British Rail, which over the past decade has absorbed four million tons of what they call PFA (pulverised fuel ash) and 1 million tons of rubbish from the London Authorities. The rubbish, mainly from West London is compacted at Brentford into enormous yellow containers, twenty feet long and eight feet high, each containing thirteen tons. These are put onto a freightliner train and taken to ARC's sidings at Appleford, where a full train of sixty containers arrives each night. At 7 a.m. each morning, a travelling gantry crane lifts the containers from the train and puts them onto ARC dumpers, which thus disposes of 800 tons of rubbish on the infilling site each day, six days a week.

My diary records:

I was shown round by Doug Raynor, the manager, and Mike Hornblow, his assistant. Their operations cover between three and four hundred acres. First they remove the topsoil and subsoil, which is stock-piled. They try and limit the number of operations by careful timing of extraction and backfilling, especially with the soil, which deteriorates while stockpiled. As the excavator moves backwards, with its conveyor, so the infilling moves forward. What happens is that the empty quarry is drained and then lined with impervious blue clay. The refuse is tipped on it in eight-foot deep bands and stamped down by a tracked mobile compactor, then a layer of inert material, then another band of rubbish with its cover – all this to ensure that the rubbish is exposed to the air for the shortest possible time. With the infilling complete the subsoil and topsoil is put back.

The land is what they called 'domed', that is contoured. It is put down to grass for three to five years, and used for hay or grazing. After that you can use it for anything you like. The rubbish backfill I saw was eighteen feet high. I also saw reclaimed land which was the local farmer's own piece – they took special trouble as it was his and 'he is really pleased'. They say the cycle from farmland to farmland can and will be reduced to a mere fourteen months. The rubbish varies according to the time of year. After Christmas it's very light – wrapping paper. In the autumn it's heavy – leaves. They have a lot of problems: noise, dust, dirt. The wind is the worst, picking up paper and blowing it across the countryside before they can stamp and cover it. They bring in Rentokill for pest control. At all costs they have to keep in with the local people and local authorities. They are learning all the time. Their chief enemies are the birds – thousands of rooks and huge flocks of seagulls that come up from the Bristol Channel. These aerial armies descend on the clay cover and dig it up to get to the rubbish beneath. Raynor and Hornblow are experimenting with gas-filled barrage balloons to keep them off – as we did with the Luftwaffe in 1940. But the most effective defence they have found so far are hunting hawks which they borrow from the nearby RAF station. These put the fear of God into the greedy seagulls.

Infilling by rubbish-disposal is only one of the ways in which ARC is seeking to repair the damage to nature which its operations necessarily involve. The object is not merely to restore the land but to make a positive and if possible imaginative contribution to its enjoyment. ARC faces two particular difficulties. One is that the industry, being low-profile, has failed to get it across how important it is, and how quickly the quality of life in a civilised society would suffer if it were denied reasonable access to its raw materials. Protesters at public inquiries refuse to see aggregate operations as anything but an anti-social activity designed to make profits; they will not recognise them as part of a public service.

Secondly, ARC is a rural-based industry operating in some of the most beautiful countryside in Britain. Objectors are numerous and vocal, and not always as high-minded as they wish to appear: they are determined to keep quarrying out of their particular patch. In 1972 a controversy over an ARC development at Batts Combe, Somerset, drove the Chairman, Gerald Mortimer, to write an exasperated letter to *The Times*: 'The Mendip Society is increasingly publicising its objections to quarrying in its area, but why does it try to portray the matter as a contest of pure altruism and aesthetic discernment on its side, against gross commercial vandalism and questionable "short-term economics" on the other? Why do its members not admit openly that they do not mind where quarrying is done as long as it is not near them, and that they could not care less who else is affected by it so long as they personally do not suffer even the most minor inconvenience?'

Whether this robust kind of counter-attack is prudent is a matter of opinion. In general ARC prefers the patient, educational approach to controversy. It is anxious to explain and to stress the positive efforts it is making. One very effective form of action is first to win the support of agricultural and conservation experts, and then to present a joint case to those in a position to influence opinion. A good example was the one-day conference at Wallingford, arranged by ARC and the Association of Agriculture in December 1984, which was attended by over a hundred teachers of geography, biology, environmental studies and agriculture. Those attending were taken round the Sutton Courtenay operations, and the conference stressed the positive opportunities presented by modern scientific extraction and landfill. Dr Stuart McRae, lecturer in soil science at London University, demonstrated that high standard restoration programmes, costing between £10,000 and £20,000 an acre, could actually increase the potential of the land.

In a crowded, prosperous country like Britain, aggregate extraction is both inevitably annoying and highly profitable, and it is only right that it should pay, as it were, a social tax. ARC's policy is to be generous but discerning. In the old days, quarry owners who got a well-justified complaint about dust from a local village would reply by giving it a tennis-court or some such amenity. It was less of a bother than putting in dust extractors, and cheaper too. Nowadays there are a great many laws governing dust and other nuisances and compliance can constitute up to thirty per cent of the costs of a new investment. But ARC is anxious to go beyond, and be seen to go beyond, mere compliance.

In fact the days when the aggregate industry resisted the pleas of environmentalists are over. Environmental costs are now automatically built into the industry's overall costs from the start, and the financial pain of compliance is no longer felt. So the old dichotomy between extractors, on the one hand, and conservationists, on the other, has vanished. Indeed, it is often people in the industry who are now the keenest and most imaginative conservers. As I pointed out at the beginning of this book, business can be, and often is, a highly creative activity. There are plenty of men in ARC who positively welcome the opportunity to make properly costed efforts to improve the environment. The best pit and quarry managers are orderly, tidy-minded people; they call it 'good housekeeping' and it nearly always goes hand in hand with efficiency. Such men take naturally to conservation and to the new art of industrial landscaping. They synchronise it with the extraction process, and the two proceed *pari passu*.

At the top there is a genuine excitement that the success of the industry is generating funds for some really imaginative ventures. As long ago as 1973, ARC went into partnership with the Game Conservancy to create a wildfowl reserve and study centre on its old gravel pits at Great Linford, on the outskirts of the new town of Milton Keynes. In 1979, after a successful partnership of seven years, ARC decided to put up a permanent building, which was completed in 1981. It has very extensive facilities: library, laboratory, conference room, offices, workshop, accommodation for research workers and, not least, an observation room for visitors with a superb view over the lakes and the great congregations of fowl – ruff,

OPPOSITE The ARC construction division was a member of the consortium which built the Mount Pleasant airport in the Falklands.

dunlin, golden plover, hobby, ringed plover, black-necked grebe, glaucous gull, spotted redshank, oyster catcher, osprey, tern, whimbrel, bittern, curlew, 145 species in all – which live and breed in the place, or visit it. There is a conservation officer, a biologist, a research assistant and a warden; the equipment is of very high standard; and the library has already built up a fine collection of specialist publications and of colour and black-and-white photographs.

The research done at Great Linford into the ecology of gravel workings, and the way in which gravel pits can be turned into first-class production habitats for wildfowl, is helping to improve the transformation of pits and quarries into wildfowl breeding-grounds all over the country. It also provides a well-equipped conference centre where small parties of opinion-formers can be shown the remarkable contribution exhausted quarries and pits can make to Britain's wildlife. The centre is the personal brainchild of Rudolph Agnew, who first conceived the idea in 1970, and it is rightly regarded as the showpiece of Gold Fields' conservation effort in Britain. It is also the focus of a wider development at Great Linford, which will turn its ten lakes into a self-financing aquatic leisure centre for Milton Keynes.

The ARC Marine 'Arco Tyne' discharging sand and gravel to form a ferry terminal at Portsmouth. The Royal Yacht *Britannia* and units of the fleet are in the background just before setting sail for the Falklands in 1982.

Personally, I find abandoned quarries the most romantic of places. That deep, overgrown marble quarry outside Syracuse where, as Thucydides tells us, the wretched survivors of the Athenian expedition were incarcerated, is to me among the most touching monuments of antiquity. But the general view is that quarries should be detoxicated and made useful; and, that being so, ARC has certainly given a lead to the industry.

Running over a hundred pits and quarries, and ensuring that its operations ultimately add to rather than detract from the environment, is only part – though it is the predominant part – of ARC's activities. It is a highly diversified group. It embodies the central principle of successful diversification: following the product theme through to the market-place. It not only extracts: it adds value to what it extracts in hundreds of different ways, and it then uses its added-value products to undertake large-scale additions and repairs to the basic infrastructure of society, both in Britain and abroad. This is the business of ARC Construction, with its head office in Sutton Courtenay, a staff of six hundred and turnover around the £70 million-a-year mark. As befits a firm which owes its distant origins to Sir Henry Maybury, it specialises in the construction and surfacing of the highest quality roads in the world. It has played a major part in the making of the M40, M27 and M11, and undertaken reconstruction and resurfacing of sections of the M1, M5 and M6. It has built motorway-standard by-passes, dual carriageways and structures for the A2, A5, A21, A40, A45 and A420.

By a logical extension, it has used its skills in manufacturing and laying surface materials to become a major contractor in the construction, extension and repair of airports. It formed a joint venture to build the £215 million Mount Pleasant jet airfield on the Falkland Islands, an operation which involved a prodigious effort to overcome logistical and engineering problems at the other end of the earth and was completed well on time in April 1985. It built the new international airport at Hodeidah in North Yemen. It has built new runways at Sumburgh in the Shetlands, Yola in Nigeria, and carried out major resurfacing and extension contracts at Heathrow, Gatwick, Colombo, East Midlands airport and many RAF stations.

Equally, it specialises in industrial surfaces and flooring, creating large-scale operating areas which are dust free, oil resistant, heavy weight-bearing, resistant to abrasion and jet-blast and capable of being laid to fine tolerances. It uses its own specialist materials or products for which it has the sole agency or license, such as Amecrete (resin-bonded granite), Gercim (a mix of concrete and asphalt), LateXfait, a special seamless plastic used for storing electrical equipment. A typical ARC Construction operation, carried out by its General Surfacing Division, was the resurfacing of the Silverstone grand prix race-track with non-skid Delugrip, of which ARC also holds the license. As part of the same technology, ARC operates heating and planning units, operating specialised machines which travel under contract to any site in Europe, and it has Shotcreting equipment for putting permanent surfaces on tunnel work. Indeed, it undertakes the tunnelling contracts itself, through one wholly owned subsidiary, and open-cast mining through another.

ARC also uses its experience in concrete manufacturing and the resources of its marine dredgers to undertake large-scale projects in sea-defence, coastal protection, civil engineering in ports and harbours and the construction of ferry terminals. To defeat the ceaseless batterings of the ocean, it builds and installs stone armour-plating and walls, stepped aprons, groynes, gabions, grouting revetments, steel sheet piling, slope protection mattresses and tetrapods. It is expert at using processed natural materials to construct protective surfaces which will stand any amount of natural or man-made aggression – no doubt, had it existed in the Middle Ages, it would have specialised in castle building. But it also carries out large-scale conventional contracts, building flats and offices, warehouses, workshops, skill centres, leisure centres, drainage schemes and sewage works.

All these operations by ARC Construction are, of course, supplied when this is appropriate by ARC's aggregates division, from its pits, quarries and

LEFT These houses were built with ARC/Powell Duffryn concrete bricks.

BELOW The ARC Leca 6 concrete block contains a foam interior which gives high insulation values.

marine dredging. But they also use supplies from ARC Concrete. This has nine strategically located factories and, like its associated company Hydro Conduit in the United States, has become the leading manufacturer of concrete pipes in Britain. Its Conbloc division is a major supplier of concrete blocks and specialised lightweight blocks, and the company makes a wide range of precast concrete products – paving, decorative walling, lintels and beams. These products, which increasingly involve entirely new composition concepts and are taking ARC to the frontiers of technology in this field, are clearly one area in which the company, in accordance with its general strategic plan, expects to expand substantially.

Visiting ARC Concrete's factories is an educational experience because, once again, it shows one the hidden machinery of the modern world – how the elements of our everyday infrastructure are actually made. It is a capital-intensive, highly automated world, employing surprisingly few people – though quite capable, in a few instances, of making use of traditional craftsmanship too. My diary records:

To Lechlade, to see the Conbloc factory and Artificial Cotswold Stone Slate. Met Chris Farren, Unit Manager, and Melvyn Bennet, Works Manager. They make high-quality building blocks. ARC builds a factory to supply a market in a given area, but this also supplies products anywhere. It makes special sizes of very high quality and the plant, which cost £1.5 million and was opened in 1982, is one of the most modern in Europe. It has a number of new technical features – not all have come off yet but will eventually. An automatic finger car takes blocks from the ovens and obviates lowering the temperature. The block factory employs only twenty-four people, including five loaders, and turns out half a million (or more) a month, up to 800,000 in summer. They make them with different colours, strengths and technical properties. Sales are worldwide. There are very high quality controls and they are constantly altering the machinery to raise them. The controls are exercised from an electronic panel which is programmed to compose and alter scores of mixes to make up multiple different batches.

At Stanton Harcourt they make lightweight blocks with foam interior. The material, Leca, is the lightest man-made aggregate in the world. Pioneering work. The blocks have cavity walls and they pump foam in. It is the first block, being 100 mm wide, which conforms to the new insulating standards. It will revolutionise traditional building, and they are now converting three other factories to make it. 'It is cheaper too, if you can sell the overall concept, but you have to convince the people at Wimpey, etc. It is perfect for the small guy building an extension,' they say, adding: 'We knew everything about concrete but this means dealing with chemicals in building materials for the first time and we are having to learn fast.'

At Lechlade they also have fifteen people turning out hand-made Cotswold tiles from processed stone. The two factories are grouped together because of the location of the planned reserve of materials (twenty-year life) and because this supply of natural aggregates is the

Laying ARC Slimline glass fibre reinforced concrete pipes, made in St Ives, Cambridgeshire.

nearest to the Cotswolds. The stuff moves from the gravel pit to the processing plant to the factory all by conveyor. Although the material for the tiles is artificially created from crushed stone, all the tiles, which come in thirty different sizes, are hand-made in moulds and then sprinkled with black and red natural stone dust; this encourages lichens and moss to grow. They are hand made both as a selling point and because it means no two tiles have exactly the same surface pattern.

Even when ARC produces hand-made tiles, however, it employs high technology in providing the composite material, and one of the primary aims of the company is to use modern skills to manufacture artificial materials from aggregates which are at once cheap and ultra-durable. Its most notable advance has been in developing Slimline technology to produce high-quality concrete pipes. These are made at its St Ives factory in Cambridgeshire and were introduced commercially in 1977. The traditional methods of making big pipes capable of withstanding heavy pressure is by using ferro-concrete reinforcement – you take huge steel coils and mould the concrete around them. Slimline pipes are made from a new patented composite – highly compacted and durable concrete reinforced by alkaline-resistant glass fibre which is strategically located, oriented and bonded throughout the concrete. In short, it is a new kind of high-strength concrete, which frees concrete pipe design from the restrictions imposed by the physical form and corrosion of steel reinforcement.

Over a hundred miles of Slimline pipes have now been laid down in the United Kingdom, in pretty well every kind of condition, and a new, bigger automated factory is under construction. Slimline has done very well in Japan. They hope to be big in Canada. Global licensing has been a long haul and has taken a lot of money and effort, but the royalties are starting to flow.

With the provision of materials and products for the basic infrastructure of society moving into various fields of high technology, ARC invests increasing sums of money in its research and technical services. Here again, I do not think most people realise how much effort goes into maintaining and if possible improving the quality of the essentials they take for granted, like the surfaces of modern roads. A growing proportion of materials used by the construction industry have to comply with the very exacting requirements laid down by the British Standards Institutions, and public bodies like the Department of Transport and the Property Services Agency have additional ones on their own. ARC experts serve on the technical committees which review and refine these standards, and at the same time strive through their laboratories and practical experiment work to improve their own products beyond the standard level and invent new ones. ARC, for instance, is working with the Transport Ministry to develop its improved bitumen for coating roadstone; and, in addition to Slimline, has developed and marketed a substance called Premix Mortar, which is quality-controlled and set-delayed, so that it eliminates the old, confusing variabilities and on-site wastage.

The central laboratories at Chipping Sodbury conduct such a multiplicity of tests and controls, using so many scientific disciplines, that it is hard for the layman to present an orderly picture. My diary records:

Shown round by Isabel Wright. She took me to what they call 'the dirty end' first. The three main types of materials they test are aggregates, bituminous materials and concrete. The lab is divided up into a whole series of separate areas where they make specific evaluations. It is a mixture of do-it-yourself rather Heath Robinson type devices, and high technology. Some on the instrumentation is very sophisticated.

They make tests of strength, abrasion and penetration. They measure hardness and softness. They examine the impact and crushing strength. They use the term 'Polished Stone Value' (PSV) and speak of 'Aggregate Abrasive Value Determination'. They measure weight loss in various conditions, using such tools as a vibrating hammer. They determine the 'Frost Heave' for bound and coated layers. Most of these are mainly for road and airport surfaces. Concrete is subjected to drying and shrinkage tests by measuring the expansion of bars and cakes of concrete. They also have attrition tests for aggregates which they need to satisfy such customers as British Rail. For BR track ballast, they need to know the attrition value when it is subjected to heavy rain, for instance. They test binder material – they want to know and measure its penetration properties. They take tests to improve tip design, to discover the factors which determine, increase or decrease moisture content. They test large quantities, such as tips, by examining drill cores. But many of the 'dirty' tests consist of subjecting the material to brutal assaults, which are carefully calculated and their effects precisely measured.

At 'the clean end' of the lab there is chemical analysis. The chemical analysis of rocks is designed to discover the chlorine content (especially in a marine context) and the sulphate contents – both important in the concrete world. But there are a lot of other chemical tests on rocks, such as limestone analysis of material used in animal feed. Industrial limestone is used for many agricultural purposes and a number of industrial ones – backing carpets, for instance. There are special tests for skid resistance, for example. Most products produced by ARC need a combination of qualities – hence the multiplicity of tests.

The environmental testing is obviously of growing importance. Noise and dust are two of the main areas. Water samples are taken and analysed, to give one instance, after complaints that too much dust is going into streams or rivers. There are tests for atmospheric pollution, produced by respirable dust, tarmacadam and other substances. The noise tests are becoming increasingly complicated and sensitive. They test for ground vibration produced by blasting, and they measure blast over-pressures, in ranges of frequencies the human ear cannot hear. Then they tape record on site. The machine breaks the noise signals into frequency components. Monitoring devices with big memory banks can be left out for months and record aural histories of sites under analysis, allowing highly accurate averages of noise levels to be worked out. They must be able to distinguish between background noises and specific

Pouring Premix ready mixed concrete.

noises requiring measurement. They use a very accurate Swiss tape recorder, whose tapes alone cost £7,000 each. The recorder is re-calibrated once a year by the manufacturer. It must be able to give precise answers which will stand up in court or at planning appeals.

Isabel Wright took me finally to their Aggregate Library. Here hundreds of samples of material are stored, numbered and classified by colour, size, texture and other criteria – materials made by ARC and by competitors. The system is made to work by carrying out regular checks and tests at all stages of the production cycle. It struck me that constant testing at every point is one of the chief hallmarks of efficiency in the world of natural resources production. It applies equally to base metal and precious metal ores and concentrates. It is something which Chipping Sodbury has in common, for instance, with Black Mountain or Mesquite, and it is a theme which runs through Gold Fields' operations all over the world.

Both at ARC's Chipping Sodbury headquarters and at Gold Fields in London there is a confident expectation that the company's expansion will continue. In London, Christopher Glynn, then a senior executive with special responsibility for ARC, told me: 'ARC will continue to grow, that's for sure – over the next five years we will spend £70 million on ARC just in the United Kingdom on improved operations. Here we will expand just on internal growth alone. ARC management is very solid and capable of looking after its own affairs.'

Such growth is not easily come by, of course. In particular, the promotion of new materials and products in the construction industry takes extra-ordinary persistence. 'Builders are very conservative,' I was told. 'They will not buy new products just because they are better. They want them cheaper too, or they need to have it proved to them that the new product is as cheap as the present one, and better too.' Selling foam-filled building blocks is a case in point. The price as well as the technology has to be right before the building industry makes any new product a best-seller. The kind of resistance ARC meets is exemplified by the fact that its sister company in the United States, Hydro Conduit, has so far declined to adopt fibre-glass reinforcement in pipe making, though they looked hard at it and even put in a pilot plant in California. Keith Orrell-Jones, who runs ARC in America but has a background in its operations, explained why to me:

> Steel is cheaper in the United States and the ratio of steel price to glass fibre is much less impressive than in the UK. We have been working to get a specification but we have difficulty in convincing people that glass is stronger than steel. It is less elastic too. If glass fibre is overloaded it does not recover, though it can take a bigger weight. So as its characteristics are radically different it needs a quite different specification. It may be that the new plant will change the cost comparison. Certainly, the plant should settle the future of Slimline one way or the other.

This kind of scepticism, from a well-disposed quarter, indicates to me that ARC's future expansion is unlikely to come from sudden and dramatic breakthroughs in new, high-technology products. On the other hand, the chief engine of growth is clearly going to be an expansion in the range of the manufactured products it sells. Glynn thinks, as he put it to me, that:

> ARC will need an additional impetus if they wish – which they do – to expand out of their existing products. The first such impetus will come from concrete tiles, a joint fifty-fifty venture they are mounting with Tarmac. Redland and Marley have got eighty per cent of the business between them. There is not room for two new entrants – so we share. Again, a hot prospect of the game in the last year or so is the brick business, which has done remarkably well in the last few years through the change in architectural fashions. So that's a possibility. With glass-fibre reinforcement, we shall see when the new plant gets going. The glass-fibre price is coming down and, though we've been making the product for eight years, this will be the first full-scale plant in the United Kingdom. The manufacturing side is still relatively small – £4–5 million in profits compared to aggregates at £40–50 million. So that's where the expansion should come.

Charles Spence is quite convinced that a greater stress on manufacturing is a sound strategy at this stage:

> We need to get deeper into products and building materials [he told me], as opposed merely to public contracting and extraction. We have to get into the manufacturing process – coating stone with cement, pipes, blocks – get into the private sector and dilute our dependency on the public sector – get into the private building sector – make concrete tiles and so forth. Public sector work poses a major problem of variable levels of demand. You put in your capital expenditure, people, structures – and then government suddenly turns the tap off. Again, there are expansion problems with aggregates. Not with labour, except for redundancy. Our labour relations are good – we are a rural-based industry. But planning permissions for sand and gravel operations, especially in the south, are becoming much more difficult. It all takes too long. Public inquiries inevitably go to appeal. Everything is moving south. Land is expensive. The concentration of population means that permission to work sand and gravel is always opposed. In any case, much of the sand and gravel is already built on. With these limitations on public work and extraction, there has been a change of policy over the past two years, and it is a big advantage to be given the freedom by Gold Fields to move more into building materials. So there we are: sixty-five per cent of our business is in the private sector, and growing; and thirty-five per cent in the public sector, and declining. This is the big fact in our lives.

The big fact at Gold Fields is that Amey Roadstone is now a major, self-sustaining money spinner, firmly set on a course of continuing growth, generating the profits for CGF which can be used in long-term projects to build future strength all over the world. The success of ARC is particularly encouraging for CGF's management because, as Rudolph Agnew says, 'We in this office put ARC together.' It illustrates the ability to generate practical and profitable ideas, which to him is the overwhelming justification of Consolidated Gold Fields' head office.

The Batts Combe quarry near Cheddar in Somerset supplies burnt lime to the steel industry in South Wales as well as aggregates to the construction industry.

7 The Central Strategy of Consolidated Gold Fields

THE TIME HAS NOW COME to examine the central strategy of Consolidated Gold Fields. The first question we have to ask is: what precise role does London play in this strategy? What is the function of the head office? Gold Fields, despite its identification in most people's minds with its South African interests, has always been a British company, based on London. When Rhodes and Rudd decided to add to their diamond interests by plunging into gold, and bought farms on the Rand, their first administrative step was for Rudd to sail to London, in November 1886, to set up a company there and raise capital. Gold Fields has been run, in a legal sense, from London, right from the start: first from 63 Queen Victoria Street, where Thomas Rudd, the brother of Charles Rudd and the Company's first chairman, had his offices; then from 2 Gresham Buildings, Basinghall Street, the offices of the firm of solicitors to which the second chairman, H.E.M.Davies belonged; then, for thirty-seven years, to Number 8, Old Jewry; then to 49 Moorgate for fifty-six years; and finally, in 1985, to 31 Charles II Street near St James's Square.

Although Gold Fields has always been a London company, financed by money raised in the City from British shareholders, the degree of supervision exercised by London has varied. Rhodes lived in South Africa and took all his decisions there. As I have already noted, he never attended a shareholders' meeting in London. Rudd lived in South Africa, too, during the time he was active in the company. He came periodically to London to address the shareholders, but as a rule it was his brother Thomas, who ran 'the London end'. In effect, Rhodes in Cape Town and Charles Rudd in Johannesburg controlled and directed the company, London merely ratifying their actions.

The position changed after the Jameson Raid, when first H.E.M.Davies and then Lord Harris, as successive chairmen, made Gold Fields in fact as well as in name a London-directed enterprise. Harris, indeed, had little connection with South Africa, apart from serving in the Boer War; he owed his eminence to his work in India and his dominance of the Kent County Cricket Club and the MCC. During his chairmanship Gold Fields became a company with worldwide interests. The day-to-day operation of the gold-mines was run from Johannesburg but policy was settled in London. Of course, it was the custom for most senior Gold Fields' staff to have served a period in the Rand. Many in fact were out there for twenty years or more. But the relationship was that of the Mother Country to a colony. London sent out one of its number to Johannesburg as 'Resident Director', rather like a governor. The company's businesses in other parts of the world were still more subordinate. As recently as the late 1950s, the boards of overseas companies in the group had to refer all decisions of any consequence to the

Gold Fields head office at 49 Moorgate, London E C2, from 1929 until September 1985.

LEFT Head office in 1892 at 8
Old Jewry in the City of
London, and BELOW the new
head office at 31 Charles II
Street, St James's Square, SW1.

A caricature of the present
Chairman's grandfather,
J. A. Agnew, who was
Chairman from 1933 until his
death in 1939.

London board. It was London which picked and promoted staff, except at the clerical level and below. London's financial control was absolute, as is indicated by this management note: 'Johannesburg has been instructed to remit to London £500,000 of their surplus funds. . . . It is recommended that Johannesburg be instructed to make a further round sum remittance of £1.5 million as soon as they are in a position to do so.' Consolidated Gold Fields, despite its global interests, was essentially a highly centralised company.

The process of devolution began in the early 1960s under the chairmanship of Sir George Harvie-Watt when structural changes were made which transformed the Johannesburg operation, over a period, into a relatively autonomous local mining finance company. Of course, once South Africa got a measure of freedom it was inevitable that Gold Fields' interests in other parts of the world would follow suit. Economic nationalism is a global phenomenon. It is by no means confined to Third World nations and coloured races. It is very strong in Australia and South Africa, which have taken both governmental and legislative steps to express it. Canada has followed suit. Up till now the United States, being a net exporter of capital, has placed comparatively few controls on foreign investment, but the import of capital is now increasing so fast that a mood of economic nationalism may well develop there too. No country, however rich, is content to see foreign-owned companies controlling its natural resources. And, since Gold Fields operates in this area, it has had to accept the principle of devolution in power and dilution in ownership for its major foreign interests.

Under the chairmanship of Rudolph Agnew, moreover, the policy of devolution is no longer reluctantly accepted as a necessity but actively pursued as a desirable end in itself. As Agnew put it to me:

> The major issue at Gold Fields in London is to recognise the realities of today and the opportunities of tomorrow and to use partnership and association as a positive force. My term is 'confederation' with us relying on shared objectives, reasoned argument and intelligence to exercise control.

He draws a powerful analogy between a worldwide investment company like Gold Fields and the British Empire. In its heyday, the Empire was accepted by the subject peoples: the fact that only 70,000 white troops garrisoned an Indian nation of 450 million people implied acceptance. Once the acceptance was withdrawn, the Empire had no choice but to devolve into a commonwealth of self-governing states. Equally, Gold Fields flourished as a centrally controlled mining investment house only so long as people overseas were prepared to allow substantial natural assets to be controlled from London. Also, only so long as London was dominant as a financial centre and Great Britain was the supplier of much of the plant needed to run mining ventures. The system, in Agnew's view, has the positive merit that it transforms the developed overseas companies (and new investments) from being mere instruments of London's will into organic centres of growth, attracting the best local talent and developing an impetus and dynamism of their own. For this purpose there is no substitute,

Rudolph Agnew, Chairman
and Group Chief Executive of
Consolidated Gold Fields.

in his view, to a locally based, publicly quoted company, whose directors
are responsible to all shareholders.

During the 1980s, confederacy has become, for all practical purposes, the
public philosophy of Consolidated Gold Fields. It has not been imple-
mented yet in all areas. Some anomalies remain. But it is defended with
great vigour by the Chairman, endorsed by the board and accepted by the
majority of executives. It is, however, by no means without its critics within
the company, at any rate at the London headquarters. The Chairman is well
aware of most of the criticism. Agnew again:

Some believe that the logic of devolution is taken further than seen elsewhere. I accept this but I believe that we are in the forefront of change: that we are pioneering concepts that will appear perfectly logical in the years ahead. In reality, I do not believe that the devolution has gone far enough: I expect to see our associates giving subsidiary management greater freedom in the future not necessarily in 'public company terms' but certainly in the running of operations.

Agnew cites other criticisms:

The confederate system is seen as severely weakening the central or parent company in a financial sense. CGF is seen as being solely reliant on a dividend flow that is under the control of associates, the passive receiver of charity from others. This is contrasted with the dynamic companies such as IBM or the Shell Group who through the control of a hundred per cent of cash flow are able to switch cash to their best advantage: to behave in the best traditions of the novelist's view of international companies. I accept that the major weakness of the confederate system is a degree of inflexibility in cash management but, with certain conditions, I do not see this as being in any way significant. I do not accept that we are passive recipients in the first place. Part of the fundamental concept of the confederation is satisfactory performance. We are not obliged to hold our investments in any single associate for ever and disinvestment may not be to the advantage of that associate. Nor is it true that we are solely dependent on dividends. As risk-takers in the first place we can raise substantial funds from the success of exploration and company form-ation. But my general answer is to counterattack such critics by asking if they have a realistic view of the mining industry and Gold Fields itself. Gold Fields is an international investor in a multitude of products the price of which is determined by world economic forces. It has to operate on a time scale that for longevity is probably only rivalled by the Jesuits. Its success depends upon its technical skill in finding the right orebodies, in the right products and in the right places and turning those into financial success. The confederacy shares the risks (and the rewards) and provides the best management motivation for people who have to have a multiple of skills and duties. The oil companies entered the mining arena in the 1970s with the same concept as that of my critics: 'control the financial ramparts' but in less than a decade most have sold their investments. They did not have the 'stomach' for the vagaries of the commodity prices and the patience demanded by the long time-scale of mining.

It seems almost inconceivable, however, that Gold Fields could now revert to the old centralised system. No one who has talked to the chairmen of GFSA or Renison, for example, could imagine them acting as a mere letter-box for orders from London. To reverse the policy would provoke a devastating internal crisis. Nor is it easy to see CGF successfully squeezing GFSA or Newmont for cash. Gold Fields' record of investing profits in the 1960s and 1970s was not impressive: the view in Johannesburg then was: 'We make the profits and London spends them – not always wisely.'

OPPOSITE The four Managing Directors of Consolidated Gold Fields: Antony Hichens ABOVE LEFT; Allen Sykes ABOVE RIGHT; Humphrey Wood BELOW LEFT; and Michael Beckett BELOW RIGHT.

ABOVE Robin Plumbridge, Chairman and Chief Executive Officer of GFSA.

BELOW Charles Spence, Chairman and Chief Executive of ARC.

To attempt to restore a kind of financial imperialism would not work. At present there are excellent relations between London, on the one hand, and Johannesburg and Sydney on the other. The various chairmen understand the new philosophy, like it and are determined to make it function. Even in the case of Newmont, a viable relationship is being established. All this would be jeopardised by a change of course.

The confederate system, then, is here to stay. That being so, and granted that London exercises authority in the form of influence rather than direct power, a number of specific questions arise. The first, and in a way the most fundamental, concerns the nature of Consolidated Gold Fields. First, is CGF simply an investment company; or is it a mining and natural resources finance house? Here we can give a straight answer: the second. CGF is not interested simply in controlling a successful investment portfolio. It believes in creative investment in well-defined specific areas: in natural resources with, in the spirit of Rhodes, as wide a spectrum as possible.

If Gold Fields is not a general investment company, how does it define and pursue its special creative interests? Rudolph Agnew explains:

The highest priority, that most likely to give the best returns, is successful exploration: Finding the right orebodies, i.e. low cost, in the right commodities; products that the world wants, in the right places, i.e. countries that are politically or fiscally stable. We do this through one hundred per cent owned exploration companies and through retained earnings in our associated companies. Orebodies are the basic material of our fortune. Today, wherever you look, Gold Fields is directly or indirectly through its associates finding new orebodies: in the Americas, in Southern Africa, in Australia and the Pacific Basin and, in the case of ARC, in Great Britain, Gold Fields is one of the most successful exploration groups in the industry. The acquisition of orebodies can also be achieved by farming-in to those found by others, i.e. GFSA acquiring Black Mountain or by investing in other successful companies such as Newmont. These routes tend to be less profitable than the first but not necessarily so. These basic activities and the development and management of the resulting operations require considerable technical and financial backup expertise. Because of the time scale I may not see the fruits of my endeavours which will be enjoyed by my successors just as I enjoy the successes of those such as my grandfather, who launched the West Wits Line. Our strategy is to make above average returns for our shareholders by investment in and management of natural resources. As these have a finite life it is a constant battle to maintain continuity of earnings, to grow those earnings and to improve the quality of such earnings.

Having defined CGF's overall strategy, let us now look more closely at the work of the London office. It was once comparatively large. It had its own technical departments. It purchased and dispatched mining equipment. It demanded, received and examined a vast amount of information from overseas subsidiaries and took decisions down to a comparatively low level. All this has now gone. The office staff is only a quarter the size. The amount of incoming paper it processes is no more than is needed to exercise strategic supervision. London does not get involved in the details of management, on the technical side or any other. What it deals in, essentially, is ideas. It is a word Rudolph Agnew uses often in relation to the London office. To him, it is ideas which make the new confederacy system work. He told me:

> The Centre, which is a term I prefer to Headquarters, can solve the problem of control to a large degree by constantly producing good ideas. This goes to the heart of what Gold Fields is about. Let me put it in a national context. I get increasingly depressed as I see our nation running down, as a nation. We have given up taking risks. We talk always about dividing the cake instead of making it bigger. Gold Fields is in the same position. If it is short of cash it must formulate more cash-generating ideas. It always comes back to ideas. The whole confederation is based on the generation of ideas.

More specifically, Antony Hichens divides the *raison d'être* or justification of the Centre into nine functions:

> First and most important, it is capable of identifying opportunity for building a business on a worldwide scale, within its particular field of expertise. The Amey Roadstone Corporation is a monument to this. Second, it finances and directs exploration. It has found two good mines in the USA. This would justify the entire exploration programme by itself – as has happened time and time again. Third, we reorganise. The setting up of Renison Goldfields of Australia was a central act. Fourth, we take major initiatives, such as the drive to acquire an interest in Newmont. It is too early to judge its success, but I think it was right in the very long term. That was entirely conceived in this office.
>
> Fifth, we raise money for all wholly owned businesses. Sixth, we provide a security umbrella against business predators. All companies are open to attack, as indeed we ourselves were and are to Anglo American. But the risk declines as the size increases. Seventh, we monitor long-term performance. Mining is a long-term business. It is never less than three and it may be fifteen years before the investment can pay off. Then of course you have a life from ten to fifty years, perhaps more. So in the earlier stages mining is very vulnerable – the results may be certain but they take a long time coming. That is why mining tends to be made up of largish groups. Eighth, there is the provision of finance for overseas associates. In the pre-1939 period it was, of course, natural to have the headquarters of an international mining group in London. From 1939 to 1979 it was, in a sense, anomalous because of Exchange Control. From 1979 onwards, with the end of Exchange Control, it has been possible to raise money in London again. Today, London is only the third largest money market in the world, after New York and Tokyo. But it is the most sophisticated general money market. So I have a feeling that our position in London will remain logical.
>
> The final point about London is that it has an international outlook.

ABOVE Gordon Parker, Chairman, President and Chief Executive Officer of Newmont.

BELOW Campbell Anderson, Managing Director and Chief Executive Officer of RGC.

That is important. Look at our spread – gold, copper, tin, zinc, lead, silver, crushed stone, sand, gravel, and so on. It is not a coincidence that the two surviving British mining finance houses, Gold Fields and RTZ, have a very broad mining spread, with widely scattered holdings all over the world, thus spreading the economic and the political risks. Some businesses have a comparably wide geographical spread, but it is concentrated in a single product, such as aluminium. We have a wide range of risk both in products and geographically.

Spreading the risks, in places and products, is thus one of London's prime functions. Countering risk, indeed, provides another justification for the confederacy philosophy, as Roy Munro, the General Manager – Finance points out:

> This office is very oriented to strategy. It is hardly involved in running things. We are locked into very large holdings and associates. This gives us a very much bigger geographical and asset spread than if we owned everything one hundred per cent. That is what impresses banks when they look at us. It also gives financial stability. So it is better to have a minority shareholding in an overseas company, especially mining.

I asked him at what point did a minority holding become a mere portfolio investment. He replied: 'Below fifteen per cent.'

Munro pointed out to me a further justification for the London Centre: formulating tax policy and executing it. His office controls the tax viability of the entire organisation. Any finance man at, say, ARC or Newmont, can phone him. He emphasises: 'Quite legitimately we can make enormous differences to our profits by getting these things right.' Tax policy was the driving force, at least at the beginning, behind the American investment policy of the 1970s. Tax is one reason why Gold Fields retains a hundred per cent control of ARC. The peculiarities of the tax laws produce distortions in international companies like CGF. In his Chairman's statement of 20 September 1984, Rudolph Agnew complained loudly about the British tax system which penalises companies earning income abroad and seeking to distribute it to shareholders:

> Advance Corporation Tax arising out of our dividend is only permitted to be offset against a portion of UK Corporation Tax and not against foreign taxes at all. Last year we paid £12 million of ACT which we were not permitted to offset against tax due to be paid or already paid on our profits. This year it is over £13 million and is a constraint on the growth of dividend payments ... This discrimination against foreign income is worrying, not only to us but also because of the increasing dependence of the UK on its overseas earnings. Mining companies have a particular problem in that they have to operate where orebodies exist, which is virtually always overseas.

Munro says: 'Tax dominates this company', and this is why he is a member of the executive committee which controls Gold Fields' strategy. 'A very big part of the Company's operations', Munro told me, 'is to make money in

OPPOSITE The No. 1 Shaft system under construction at the Leeudoorn mine on the West Wits Line scheduled to produce some 8 tonnes of gold annually by 1995.

To reduce overall production costs, the Magma copper mine in Nevada has introduced heap leaching of oxide ores to complement its underground mining production. Plans have been announced to create a new company comprising Magma and Pinto Valley which will be established in early 1987. Newmont will retain a 15% interest; the remaining shares being distributed among Newmont's shareholders and the management of the new company.

one way or another by taking advantage of the law.' Agnew has laid down company policy firmly: 'We never do anything for tax reasons. But everything we do, we have to get the profit-angle right, and that is tax-dominated.'

There are, then, a number of specialist tasks which the Centre performs, but in essence it is a place where strategy is planned and ideas hatched and discussed. That means it is a place for argument. For a company of its size, CGF head office is not large but it contains a notable number of highly intelligent men and women who probe and analyse company policies. In some ways it is like a Think Tank. Like it or not, it must engage in that inchoate science, futurology. The lead-times in mining have always been long and they are getting longer. Sinking a deep shaft system in South Africa now takes fifteen years. The whole time frame is changing mainly because the depth at which ore is worked is getting deeper. In quarrying, lead-times are being added to by the growing maze of central and local government regulations: they used to say it took about eighteen months between negotiating to buy a field and digging the stuff up. By 1974 it was four years. Now it is even longer. So the future is scrutinised carefully and rival theories canvassed. This is not a tub-thumping argument. The style is highly cerebral, rather like rival chess players, thinking a dozen moves ahead.

In addition to the continuing financial debate, I have identified some crucial areas of discussion which constitute what I would call the chief strategic problems facing the company. The first concerns geographical balance, and in particular how to get the right mix of advanced countries which enjoy political stability and the benefits of English-type legal systems, and other less stable countries, where orebodies may be richer, costs lower and opportunities correspondingly greater. The choice is not as wide as outsiders might think. As one Gold Fields executive put it to me: 'You are at the mercy of the Almighty – it's where God put the minerals in the first place. Traditionally we went into English countries where they spoke English and had English law, as parts of the British Empire. Now we tend to go into countries where the mineral is.'

In some ways, of course, a mining company is peculiarly vulnerable to the consequences of political instability – particularly expropriation. 'In a typical mining project', Peter Fells says, 'about eighty per cent of capital expenditure has taken place before the mine reaches full production. It is a very tempting target for an unscrupulous government. It is very tempting for an unsophisticated government too, because it sees huge profits being made and forgets the huge investment earlier. It isn't only expropriation which you fear. The history of mining is littered with other examples – penal taxes, export taxes and so on.' That is why Gold Fields pays almost as much attention to political exploration as to mineralogical exploration. On the other hand, among international mining companies the fear of punitive or foolish actions by Third World governments is less than it was in the 1960s and 1970s. 'Mining assets are in some ways impervious to human folly, unlike factories which can be burnt down overnight.'

Louise du Boulay, former editor of the Gold Fields' research publication, *Gold Survey*, stresses the fundamental truth: 'If we're going to be a mining company we have to operate in tricky areas. The United States is safe, but then it's a high cost mining country.' The political difficulties found in Third World countries are balanced by a growing number of legal, environmental and social difficulties in the advanced countries. As we have already noted, there are very severe restraints on mining in North America and Australasia. The geographical choices, therefore, are often finely balanced, in terms of the likely long range returns on investment. Gold Fields can and does seek to clarify and quantify the arguments by putting the political as well as the cost factors into the computer. But all the computer really tells you in the end is that mining anywhere is a pretty risky business – which you knew to begin with. What the figures do seem to indicate is that it is in some ways more risky than ever. As Louise du Boulay warns: 'It's not as easy as it was to make money in mining. Many companies are still run by men brought up at a time when making money in mining was easy – and that's their problem.' It is certainly not Gold Fields' problem. No one at its centre has any illusions about easy profits. There is constant stress on the inevitability of risks and the need to think very far ahead for results. 'It is sometimes wrong', Louise du Boulay notes, 'to pull out after reverses, when you've already paid for the learning-curve.' Rudolph Agnew insists: 'Mining is an industry for cool nerves and long-term horizons.'

Research is vital to the ARC product development programme.

In addition to the geographical spread there is the product spread. CGF's next problem is: what is the right combination of minerals? What is the right combination of minerals and other products? Up to a point, at least, CGF wants as wide a spread of minerals as possible. This raises the question of how much it ought to spend on exploration. It is at present spending a great deal, in South Africa, in North America and in Australia; rather less in other areas. Mining houses which finance large and costly exploration efforts tend to feel a glow of moral righteousness. Like the wise virgins, they are thinking of the future. Expressions like 'bottom drawer' are used with considerable satisfaction, in Johannesburg, in Denver, in Sydney.

The history of the mining industry teaches that a scientifically conducted and adequately financed exploration programme nearly always brings results in the end. A geologist told me: 'One find will justify a professional lifetime'; equally, one mine will more than pay for a big exploration programme over many years. A mining company is, or ought to be, a high-spirited animal, and the whiff of discovery brings a flaring of the nostrils and a huge increase of morale. Buying assets is just a financial transaction.

No doubt the solution is the one that Gold Fields has pursued: exploring and buying. By exploring it has found good mines in the USA; by buying it has acquired an interest in two more found by Newmont. To get the mineral mix right both methods have to be used, and even then a company can find itself too 'coppery' or 'tinny'. To the dilemmas of the mix the trade cycle adds a third and related problem. As in recent years, a severe depression in the trade cycle may mean low prices for virtually all base metals. A wide spread helps even in this case – in Australia, for instance, mineral sands prices, corresponding to the manufacturing trade cycle, have picked up while copper and tin remain low. But as a rule, base metals have to be balanced by products which are less responsive to the cycle, are acyclical or best of all counter-cyclical. To add to the problem, mining companies with their long lead-times, find it hard to adjust to the cycle. As Peter Fells points out, 'for deep mines it takes five to ten years to dig down, and then another two to get into production – so ten years, say. The consequence is that you have to predict markets over a long period. Because of the long gestation period, you know about capacity which will come on-stream over the next three years but thereafter it becomes very speculative. This is why surveys are so wild. People over invest in capacity. In the early 1970s there was a huge boom in nickel mining. We were criticised for not having any. But it was clear there was going to be a surplus. Molybdenum is the same.'

Gold Fields now has as wide a spread of minerals and other natural resources as any other company in the world and that is undoubtedly a protection. But a full insurance against the trade cycle would almost certainly carry CGF into areas where it does not want to be. The debate, in particular, raises again the vexed question of a manufacturing sector. The theoretical arguments for CGF having a major manufacturing sector are strong. Many Gold Fields' executives accept them. Others are sombrely aware of the risks a purely mining strategy involves especially if the current depression in the industry continues for many years.

My guess is that in the long term, because the advantages of a manufacturing/mining balance are so strong in principle, CGF will again seek to acquire a manufacturing arm. In a sense, management strategies are as much fashion as rational calculations. From time to time they alter, just like hemlines, not from any deeply based calculation but chiefly because of the human need to change, to be seen doing something new. But that is to look well into the future. For the rest of the 1980s CGF is committed to a non-manufacturing strategy.

In any case, many senior CGF people, in London and abroad, have more fundamental objections to manufacturing, especially as it relates to management. Roy Munro, for instance, strongly defends the decision to sell off the remaining manufacturing interests in the United States: 'I personally believe we should get out. Eventually something will go wrong. The time to get out is now, while it's doing well. Some people say our portfolio of mining assets is a low-quality investment, but I don't agree. Over the years they have been our most stable source of income. If one of our mines gets into trouble we have competent people to put in and get it right. But if our steel mill gets into trouble we have no one to put in – and sooner or later a manufacturing business always gets into trouble.'

Some of the most formidable people in Gold Fields believe the management point is absolutely decisive. Robin Plumbridge told me:

> You seek to invest in those areas where you get the highest return, and these are bound to be related to those areas where you have the management skills. Our management has grown up in an environment where the perspective for decision making is long in time – fifteen years. Our ability to take good decisions which have medium or short-term impact is limited. In manufacturing, the so-called long-term perspective is only three to five years, while ours is fifteen years. People who mature

LEFT A share certificate; the first issued by Gold Fields in 1887.

BELOW A random selection of share certificates which represent the spread of natural resource investments which Gold Fields maintains in London as an important and very profitable dealing portfolio. These interests are in addition to the long-term strategic holdings in Group companies.

Ryan Consolidated is a joint venture between Ryan International and Consolidated Gold Fields. The joint venture was formed in 1986 and recovers coal from old colliery waste dumps. Here, in the valley, a worked out site has been contoured and planted.

under the mining style of management don't transplant easily to manufacturing, and vice versa. The time frame is vital. It produces quite a different type of management.

However, the debate over manufacturing will never finally be ended; and in the meantime, Gold Fields' substantial and increasing commitment to construction materials represents a compromise of a sort. The construction industry is itself cyclical, of course, but it is by no means the same cycle as base metals. Hence ARC in Britain and the United States helps to give a much-needed balance to CGF's overall spread of interests. It is true that ARC, particularly in Britain, is now moving heavily into the manufacturing of construction materials. But this is a pseudo-contradiction rather than a genuine one. Companies which quarry aggregates move naturally into ready mix concrete and then into concrete products, just as they move naturally into construction work itself. Management is interchangeable. Beyond a certain size, it is a normal progression. The ARC people know the business of making pipes very well, and so far as the style of management is concerned, they are well equipped to penetrate the manufactured materials side much further. No one in Gold Fields, let alone in ARC, has any doubts on this point.

What also surprises is that CGF's involvement in construction materials is so far virtually confined to Europe and North America. But this touches on another strategic problem. Should Gold Fields be just a mining finance house or should it embody the much broader concept of a natural resources finance house? There is no doubt where CGF in London stands on this issue. Indeed, granted the enormous success of its own brainchild, ARC, how could it do otherwise? Under Rudolph Agnew, Gold Fields has become wholeheartedly and irreversibly committed to the extraction and benefici-ation of the full spectrum of natural resources, and now that its stone, sand and gravel company is its biggest money spinner, it is quite inaccurate to call it just a mining finance house. In its American operations, the broader definition is still more applicable, for ARC and particularly Hydro Conduit are the soundest elements, both as current performers and in terms of future promise. But it is a different matter in South Africa and Australia. In Johannesburg, Robin Plumbridge is adamant on this point: 'Our definition is the narrower one of mining and beneficiation. I doubt if we would want to get into construction.'

Granted his strong views on the long-term style of mining management, this is logical. GFSA has spread into the energy field, but chiefly in coal, which of course is mining too; it is not interested in oil, except as an invest-ment in Sasol. Renison in Australia is certainly interested in oil, which it is pursuing with great energy. But otherwise it is much closer to the South African philosophy than to London's.

The contradiction, then, in Gold Fields' product strategy is not that it is involved in construction materials with a large and growing manufacturing element, but rather that two of its principal overseas associates have not moved into construction materials at all. The anomaly is particularly marked in Australia, where Renison's most successful division at present, its mineral sands operation, is particularly well equipped to deal in aggregates and similar products. In many ways the mineral sands business is closer to quarrying than to mining proper. A big restructuring is taking place in the Australian aggregates business, suggesting it is ripe for penetration, and a move into this area is being debated in Renison's Sydney headquarters. In the long run, it is not hard to see an ARC-type operation in all four of Gold Fields' principal geographical theatres, and when (and if) this happens, the natural resources philosophy will have triumphed completely.

A further strategic problem confronts Gold Fields, and it is linked to its own creed of devolution. Has Gold Fields de-colonised its subsidiaries only to find itself colonised in turn? Can it cope with the Anglo American factor? Minorco, a company within the Anglo American Group, holds twenty-eight per cent of Gold Fields' stock. This stake was built up in 1980 and topped out in a notorious dawn raid that led to a change in Stock Exchange regulations. When I asked Agnew about the Minorco factor, he replied:

There has been considerable speculation concerning the reasons for the 1980 operation and Minorco's future plans. I accept the reasons given at the time that it was intended to protect the 'controlled' independence of GFSA by ensuring the independence of CGF. During the 1970s CGF was ambivalent about its investments in South Africa and speculation grew as to whether CGF would disinvest from GFSA. GFSA is a crucial factor in the South African mining industry, especially in gold. The Group's gold

The Gold Fields Environment Trust finances conservation projects and training schemes for the young unemployed in rural areas. Here members of a YTS course are learning to build dry stone walls at the Coomb farmhouse countryside training centre, near Dorking in Surrey.

production is the lowest cost and is growing as new mines are developed. Compared with GFSA, the two major rival groups, Anglo American and Gencor, look middle-aged. GFSA is the swing producer between these two groups; the plum which neither would wish to go to the other. Also the plum in which CGF appeared to be losing interest. However, by the late 1970s CGF ambivalence disappeared with clear statements that we understood the value of the GFSA investment and intended to stay with it. Shortly afterwards rumours grew that someone, possibly Gencor, was building a stake in CGF presumably with the objective of gaining control of GFSA. Anglo reacted to these rumours characteristically by moving at speed to take a blocking shareholding in CGF.

Then followed a joint declaration between our then Chairman, Lord Erroll of Hale, and Harry Oppenheimer, that there was no intention Anglo would go above a thirty per cent holding in our company, and that the two groups would remain competitive. We accepted two Anglo directors on our board and over the last four years all the evidence is that the communiqué is being stuck to. They have not tried to influence the direction of CGF in favour of Anglo. What they have done is to ensure that their position as a shareholder is not being harmed.

Agnew's relaxed attitude towards the Anglo American factor makes sense, but it would be idle to suppose that this big and potentially predatory holding does not make a difference. Probably the difference seems greater in Johannesburg, where the implicit threat is nearer and more obvious. In London, the Anglo American factor is felt more as a restraint on policy, though not necessarily a malevolent one. CGF's relations with Anglo, rather like Newmont's with CGF, have now moved from outright hostility and intense suspicion to what Hichens calls the constructive phase. But suspicion remains. 'Anglo', says Hichens, 'has a reputation of emasculating previously great companies by trying to interfere too much.' Hence it is important that 'they are not participants in our management or controllers of it.'

The shadow of Anglo American over Gold Fields, its determination not to be outstripped by Gencor, the notion of Gold Fields as a swing-company in the struggle – all point to the continuing importance of gold. Gold is the historical, professional and financial core of the company. It is also its emotional core. Antony Hichens told me: 'If we correctly judged gold was going downhill for ever I hope we'd have the moral courage to get out. But,' he added quickly, 'we think broadly the opposite.' In defining the product philosophy of CGF, he stressed: 'The thread we want to pursue is mining and natural resources, and in particular gold.'

The gold price is notoriously difficult to predict, both in the short term and the long term. But it would need a bold man to predict there is no future for gold in view of its spectacular performance in the late 1970s. Gold is no longer the official basis of the world's monetary system. In many ways it is like any other commodity. F.D.Roosevelt fixed the gold price at $35 an ounce in the mid-1930s and that remained the only legal price for nearly a quarter of a century. In 1958 the Reserve bank of South Africa began to sell gold to private persons, and thereafter there was a fixed-price market at $35 and a free market price which edged slowly upwards. In 1972 the American government raised the official price to $38 and the following year to $42.22; on 1 January 1975 it abolished the law which forbade its citizens to own and trade in gold. The final abolition of the official price came in 1978, when the IMF rules were changed to allow Central Banks to buy gold at any price. Freeing gold added considerably to its value but turned it into a highly volatile commodity. It has tended to be a mirror image of the dollar, climbing when the dollar weakens, falling when the dollar is strong. It rose to over $800 an ounce in 1980, then drifted downwards when the dollar recovered. In 1985, with the dollar still strong but inclined to weaken, gold was in the $300–400 band.

The effect of these huge swings on the fortunes and thinking of a company like Gold Fields has been dramatic. In one respect CGF can remain impervious to the volatilities of the market because its main investments are in the lowest cost gold-mines in the world which can make a profit mining gold at almost any foreseeable price. But Gold Fields also has to think about creating the mines of the 1990s or even the 2000s; and, directly or indirectly, it has a lot of money invested in more marginal gold-mines in the United States and Australasia. It must also balance the likely price of gold against political factors. With gold at $800 an ounce, it looks absurd not to invest in South African gold-mining; at $300 an ounce, the political risks seem more formidable. There is also the question of what other people are doing. The spectacle of gold hitting the magic figure of $800 an ounce inspired a gold-rush all over the world. With the long lead times, new mines are coming on stream, or projects nearing completion, after gold has fallen from its peak. In 1984 for instance the gold price fell from a high of $405 in March to $321 in mid-December. Nevertheless, the number of new gold projects rose during the year from 108, valued at $8.73 billion, to 137, valued at $9.77 billion. Not all these embryonic mines will come into existence. But many will, increasing the supply pressure at the low-grade end. The flight of mining capital into gold projects reflects disillusionment at the continuing low price of most base metals. Gold, to put it in a negative sense, is now the least unattractive metal to people who want to venture their money.

Of all the mining finance houses, Gold Fields is perhaps the best equipped to look at the future of gold and has the best record in doing so perceptively. Like other firms, it has a big commodity department which monitors the markets in which the company operates, forecasts supply and demand and therefore price, and so helps to get production costs right and project the cash flow. It spends more time and money looking at the gold market than any other. Its *Gold Survey*, first published in 1968, was the earliest, and it is still the leading publication in the field, being primarily a historical study of the past year in the world's gold markets.

Louise du Boulay, former editor of *Gold Survey*, stresses: 'This publication is virtually the only set of figures on supply and demand of gold.' It is compiled by extensive research on both sides of the industry, particularly the demand side: 'We visit dealers, jewellers, even illicit dealers.

Gold Fields has presented gold and silver medals annually since 1912, which are awarded by the Institution of Mining and Metallurgy in recognition of outstanding technical contributions reported in the transactions of the Institution.

GFMC continues to concentrate its exploration programme in North America on finding gold. Three promising sites are currently under evaluation, of which at least one could well prove, in 1987, to be economically viable.

We don't trade – we are only a producer. So that's why people talk to us.' The publication does not actually forecast what gold is going to do. Essentially it is a by-product of the commodity department, which of course has to make forecasts, about gold and other metals, for internal consumption. Sam Brooks, the manager then responsible for mining services, described to me the department's objects and methodology:

Short-term predictions, that is one to two years, are not very important in their effects on production. But the long term, by which I mean ten years or more, is vital. Should we be looking for gold-mines? We are concerned chiefly with gold, but we do the same analysis for copper and tin on a regular basis, and slightly more irregularly for lead and zinc. We can do it for other metals too – we're about to do a one-off study of molybdenum. We are supposed to keep a general eye on all commodities. We think it's important not to be obsessed by price watching. You must develop the feeling – is a crunch coming upon the supply side and so on. This was how we first got a hunch on gold in the late 1960s – demand rising and supply static. So we correctly forecast a substantial rise in the price.

Historians have the best training and background for this kind of research, but they must have numeracy. Economists tend to be obsessed by cycles. Engineers find it difficult to fill in gaps where information isn't available – they want to be able to go out and measure things. Historians have feelings of intuition and imagination. You mustn't be obsessed with minor details but keep a sense of proportion. You need the ability to synthesise a fairly large mass of incomplete data, bring it together and paint a picture. The prime requirement is curiosity – you must want to know why things are happening.

How do we set about it? There is a great deal of desk research. You add up all the mines in the world. You visit them. You visit Departments of Mines. You go to the big companies. What do they think small-scale workers, who don't publish accounts, are producing? Then the demand side. You trot around to the jewellery-producing centres. How much are they selling? Who to? The industrial usage is straightforward. You talk to refiners. People are forthcoming if they know you'll keep it confidential. We have established a first-class reputation for this. Much of our work is akin to investigative journalism. Lots of facts are known, but we are bringing them all together to produce something new.

The really interesting element is investment buying. Are the gnomes buying gold? Are the Russians selling? So you visit bullion dealers in Zurich, New York, London and other centres. You must have a large range of contacts and talk to them regularly. You visit the major financial centres three times a year, then take a swing for two or three months through Japan, Korea, Australia, Latin America, combining visits to banks, miners, Treasuries and so forth. All these people are helpful provided they know you won't take advantage. And it's so much easier when specialist talks to specialist. It's a small world. At a gold conference, with two to three hundred attending, you will have all the big players in the game.

The political side is the most important and affects every commodity. We are sceptical about our forecast because of the unforeseen factors of which the political is the most likely. But long-term forecasting is still worth doing. The trends tend to be more stable than the numbers. A particular event has to be very remarkable to reverse a definite trend. We had a strong view gold would rise, so we looked at particular projects – not to start production but to keep in the bottom drawer. It is much trickier to get the timing right. You have to guess: when will the pressure produce the desired result (especially in politics)?

Through the commodities department the men who run Gold Fields get all the best information available to help them reach decisions about future trends in markets. The commodities put the stress on the long term, and that is right. Indeed, it is the public philosophy of Gold Fields, which thinks not from year to year but in decades. This philosophy was one main factor in the move from 49 Moorgate in the City to 31 Charles II Street, St James's Square, in the West End, which took place in the late summer of 1985. The Moorgate building was a famous, or notorious one, built by the Soviet trading firm, Arcos Ltd, in the early 1920s, and in fact used as a centre of espionage. They packed 1,000 people into it and built enormous strong-rooms. On 12 May 1927, on the order of the Home Secretary, 150 police surrounded the building, entered and searched it, and used oxyacetylene torches to open the strong-rooms. The trade mission withdrew and Gold Fields bought the building for £140,000 in 1928. The strong-rooms were henceforth used for storing the company's records. For nearly a hundred years Gold Fields has operated from the City. From the days of Cecil Rhodes, whose reputation dazzled City brokers and ordinary shareholders alike, Gold Fields has always been highly successful in raising City money.

In some ways the City is still an excellent location for a company like Gold Fields. Peter Roe, the recently retired company secretary, told me:

> Our main business after the war was raising the money to bring the new South African mines into production, so proximity to the City and City contacts were useful. But today we no longer want to be so financially orientated. The City is a great place to work in – you can get your advisers together quickly and do a deal. But you are right next to the market and your analysts tend to hear every market rumour and believe it. I thought for a long time we ought to move out of the City, and Rudolph Agnew definitely prefers the West End. He thinks the City breeds too short-term a view, and he likes the long view. Agnew believes the West End has a more vivacious atmosphere and he thinks the culture shock will do us all a lot of good.

The new West End headquarters is basically an Adam house with an extension by Sir John Soane, altered by the architects Edmeston & Gabriel in 1911. It provides an impressive and elegant setting for Gold Fields' policy making during the second century of its existence. This will be the place where the ideas will be produced, for as Rudolph Agnew says, CGF lives on ideas at the centre: 'Most of our problems', he emphasises, 'lie right in this

Colin Fenton, Managing Director of the Northam Platinum project in South Africa turns the first sod at the site of the No. 1 shaft: a traditional inauguration ceremony for all new mines. Northam, scheduled to begin production in 1992, has reserves for many decades and working costs which compare favourably with competitors.

office, and nowhere else.' Charles II Street will be the forum for hard-headed, creative thinking about natural resources all over the world. But this is not to say that Consolidated Gold Fields, in its second hundred years, intends to be over-deliberative. It must think in the long term. But it must also embody the spirit of adventure and the appetite for risk taking.

That, indeed, is the ethos of the company. 'We have always had a buccaneering outlook', Antony Hichens says:

> We have a go. We take risks. We look for new things to do. Our top people are encouraged to take risks and not normally punished if the risks do not come off. This attitude is sharply different from more orderly and logical companies. It may even lead to a certain amount of confusion. When I first joined the company I found it hard to find out what they were doing, because the willingness to accept risks is hard to grasp as a company strategy, and it certainly can't be put down in a formal strategy paper. But you learn by example. We're enormously proud of the West Wits Line – but the risks we took were breathtaking. Again, we're enormously proud of Amey Roadstone. But this happened because the company is used to having a go, and executives knew their careers would not be blighted if there was a failure.

No one expresses this ethos of adventure and risk better than Rudolph Agnew himself. 'I am going for growth in assets and earnings,' he told me:

> These will largely depend on successful exploration and good acqui-sitions. Because we are placing so much emphasis on exploration, we will go increasingly to the great mineral provinces of the world – to the Americas, to South Africa, to the South Pacific, to Australasia. The

Part of the Waruwari range, site of the Porgera joint venture in Papua New Guinea. RGC has a one-third interest in this exciting gold prospect which has reached an advanced stage of evaluation, and could be producing some 15–20 tons of gold a year by the early 1990s.

means we will employ will be the partnership between Consolidated Gold Fields and its federated associates. There are sirens seeking to distract us, the short-term considerations that companies are increasingly subjected to – earnings per share league tables, analysts' reviews and so on. I lump them together under the general heading of vanities: Chief Executives hate to see their company described as 'dull'.

But my starting point is the romantic nineteenth-century attitude. The mining industry is a great adventure. Sure, it should be highly profitable. But it is not a game susceptible to stockbrokers' charts because of its very nature. You have to cope with great political pressures. You need good nerves, a long-term view. It is an industry which is very involved with people – including the sheer physical safety of people. Our founders, Cecil Rhodes and Charles Rudd, epitomise the typical dichotomy of mining. On the one hand you have the ruthless merchant-adventurer type, on the other the conservative professional needed to support the lead given by the adventurous. The good mining finance companies are the ones that blend these two streams successfully. The blend is not unique to CGF. Val Duncan of RTZ and the Oppenheimers of Anglo American are other examples. To me, Gold Fields is a great story of success and failure – and success again. It reflects the genius of the British race, which is a genius for merchant adventuring in obscure parts of the world. The question, if you agree with my romantic vision, is whether Gold Fields lives up to its tradition of adventure, or whether we're faceless men working solely for our pensions.

8 Moving into the Twenty-First Century

Mining and mineralogy together form one of the principal dynamics of civilisation. From the beginning of recorded history, the quest for precious and useful substances buried in the ground has led men to explore and settle in some of the most remote and difficult places on earth. The ancient Egyptians, searching for silver, for gold and jewels, for the granite and other hard stones they carved so skilfully, opened up all of north-east Africa to trade and letters. Primitive smiths, working first in copper, then bronze, then iron, punctuated the spread of urban culture with their hammers. It was Cornish tin which first brought sophisticated strangers to Britain, and the gold of South Wales which attracted the Romans.

The lust for gold, above all, is the force which drives men to penetrate hitherto inaccessible territories. It is amazing what men will do and suffer to get it. It opened up central and southern America. It drove men across the Rockies to California, and up the Yukon into Alaska. Diamonds, and then gold, transformed South Africa into a modern industrial community. In Australia gold caused multitudes of almost penniless men, pushing their belongings in wheelbarrows, to trudge thousands of miles along the coastline until they came inland to Kalgoorlie.

Mining discoveries are the great precursors of infrastructures. The Romans built roads and ports to make their mines more accessible. The Spaniards set up the first regular fleet convoys across the Atlantic to ensure the safe delivery of their gold and silver bullion. The railways from Johannesburg to Cape Town and the East Coast were the direct result of the Rand discoveries. It was the gold of New Guinea which created the world's first regular air-freight service in the 1930s. The process continues. The beautiful but harsh north west of Australia is now being provided with ports and roads and railways because of the great iron discoveries there. Thus the frameworks of advanced societies come into being and are taken for granted, and often their origins in mining are quickly forgotten. A mineral sands engineer said to me: 'We discover a deposit on a remote beach. We build a road. We put in water, electricity and sewage. Then people make use of our road and start to build houses near the beach. A town evolves. Almost the first thing they do is set up a society to protect their community from the environmental effects of mining. They forget that, but for us, there would be no community at all. So they stop us mining and we go on to the next place.'

That is the essence of mining. It is a pioneering activity. It pushes forward the frontiers of civilisation, and when settled communities catch up, it moves on again. It is highly mobile. It is also cosmopolitan. Most of the skilled men I met on my travels, whether geologists or mineralogists or mining engineers, had been all over the world, and worked in many

Drilling at the GFMC Chimney Creek prospect in Nevada, now scheduled to begin gold production at the rate of 4–5 tons a year early in 1988.

different kinds of mines and countries. They have an international outlook. They go for the broad horizon.

Until the last decades of the nineteenth century, mining remained a fairly small-scale affair. Mining communities were small. With the advent of industrial capitalism, the scale of mining activities increased steadily. But it was not until the great Rand discoveries of the 1880s that the full potentialities of large-scale industrial mining began to be explored. Gold Fields was born in this epoch and its first century, coinciding with the second century of industrial capitalism, has seen mining evolve into a highly sophisticated business, employing enormous financial resources and the very latest technologies. It has become a huge creative force in the world, as indeed Cecil Rhodes foresaw.

We are now about to enter the third century of industrial capitalism, as Gold Fields completes its first. This follows a prolonged and in some ways intense recession. In the second half of the 1970s, it was fashionable to predict the end of capitalism. The world was said to be running out of resources. Such dismal bodies as the 'Club of Rome' predicted inevitable stagnation and decline. 'Zero growth' was a cant term. No one involved in the actual business of recovering natural resources from the ground believed in these Jeremiahs. Miners develop a professional consciousness of the inexhaustible riches of the planet. They know that, so far, we have been only scratching its surface. Miners are optimists. That is what drives them forward, up into the hills, down into the bowels of the earth.

Now the wave of pessimism, so characteristic of the 1970s, has ended, to be succeeded by a new wave of realism. The future looks reasonably propitious. The world has learnt a lot of valuable experience from the illusions and follies of the 1960s, the despair of the 1970s. The world economy is recovering. The revolution in information technology is gathering pace. Some time in the 1990s it is expected to be married to the revolution in biotechnology, now still in its early stages. The result will introduce the twenty-first century with an explosion of new products and service industries. At the same time, the so-called developing countries in the world, including China, one billion strong, and India, nearly as numerous, will be becoming major industrial nations with rising living standards, consuming natural resources in quantities we now find difficult to imagine.

Consolidated Gold Fields and its confederation of associated companies are well placed to play a part in these tremendous developments. They have an exceptionally wide and varied spread of reserves, in minerals, aggregates and energy. They are actively looking for more, all over the world. They have kept in the forefront of technology over the whole range of natural resources provision. From their mines and quarries, present and future, they can provide their full share of the metals and materials which this great global surge forward will demand. Perhaps most important of all, Consolidated Gold Fields has kept alive its spirit of adventure, and imbued with it the many thousands of skilled men and women who share and delight in the company's expert task of wealth creation for the benefit of humanity.

Acknowledgements

The works I have found valuable while preparing this book include:
A.P.Cartwright: *Gold Paved the Way* (London 1967), and his study of the West Driefontein flood crisis, *Ordeal by Water* (Johannesburg n.d.); another study of Consolidated Gold Fields, Harry Hake's *Variety: the Very Spice of Life* (unpublished); another unpublished study, this time of the Amey Roadstone Corporation: *History of ARC – Forty Years On*; Robert H.Ramsey: *Men and Mines of Newmont: a Fifty-Year History* (New York 1973); Monica Wilson and Leonard Thompson (eds): *The Oxford History of South Africa*, 2 vols (Oxford 1969–71); D.M.Schreuder: *The Scramble for Southern Africa, 1877–1895* (Cambridge 1980); Freda Troup: *South Africa: An Historical Introduction* (London 1972); John Marlow: *Cecil Rhodes: the Anatomy of Empire* (London 1972); John Flint: *Cecil Rhodes* (London 1976); Basil Williams: *Cecil Rhodes* (London 1921); Duncan Innes: *Anglo American and the Rise of Modern South Africa* (Johannesburg 1984); H.S.Frankel: *Capital Investment in Africa: its Causes and Effects* (Oxford 1938); Geoffrey Blainey: *The Peaks of Lyell* (Melbourne, 4th ed. 1978), and the same author's *The Rush that Never Ended: a History of Australian Mining* (Melbourne, 3rd ed. 1977); I.W.Morley: *Black Sands: a History of the Mineral Sand Mining Industry in Eastern Australia* (Queensland 1981); R.J.Adamson (ed.): *Gold Metallurgy in South Africa* (Johannesburg 1983).

Illustration Acknowledgements

The publishers are grateful to the following for permission to reproduce their photographs:

De Beers Consolidated Mines 26, 27;
Guildhall Library 229;
Illustrated London News 26, 27;
McClung Collection – Lawson McGhee Library, Knoxville 117;
Montana Historical Society, Helena 138.

All other illustrations are from Consolidated Gold Fields
Picture Research by Anne-Marie Ehrlich.

Index

Page numbers in *italics* refer to illustrations and captions

African Estates Agency, 28
African Gold Investment
  Company, 28
African National Congress, 93
Aggeneys Farm, 98, 100
Agnew, J.A., 41, 44, 45, *230*
Agnew, John, 158
Agnew, Rudolph, 53, 109, 156,
  159, 202, *204*, 205, 227,
  230–9, *231*, 243, 247
Albu, G. & L., 28
Alder Gulch, 137, *138*
Allen, Paul, 181
Alumasc Ltd., 52
Amalgamated Roadstone
  Corporation, 202–3
American Girl, 114
American Standard, 106
American Zinc, 104, 106, *117*
American Zinc, Lead and
  Smelting Company, 51, 52
Amey, R.W., *200*, 203, 204,
  204
Amey, William, 204, *204*
Amey Roadstone Corporation
  (ARC), 53, 54, 122–7,
  201–27, 234, 235, 239, 242,
  248
Amey Roadstone Corporation
  America, 118, 122–9, *124*,
  127, 242
Amoco, 164
Anderson, Campbell, 199, 235
Anglo American, 96, 139, 155,
  244
Anglo American Corporation
  of South Africa, 48–50
Anglo American Gold
  Investment Company, 64
Anglo-French, 96
Anglo-French Exploration
  Company, 50
Anglo-Rhodesian, 141
Anglo Transvaal, 49
Anglo vaal, 48, 63
Angola, 92
Apex Mines, 50, 94, *95*, 96, 97,
  98
Appleford, 118, 122–9, *124*,
  127, 215, 242
ARC, 53, 54, 122–7, 201–27,
  234, 235, 239, 242, 248
ARC America, 118, 122–9,
  124, *127*, 242
ARC Concrete, 222
ARC Construction, 220–1
ARC Marine, 213–14, *214*
ARC Premix, *202*
Arcos Ltd., 247
Argentine, 110
Arizona, 125
Arnold, John, Ltd., 203
Asarco, *117*
Associated Minerals
  Consolidated Ltd. (AMC),
  159, 188–94, *189*, *195*

Associated Sand and Gravel,
  124, 125, 126
Association of Agriculture, 217
Aswegen, Hennie van, 100
Atherstone, W. G., 20
Augrabies Falls, 100
Australia, 158–99
Australian Copper
  Development Association,
  173–5
Ayer, Charles, 141
Ayre, Mike, 169, 171, 175

Bailey, Abe, 45
Bailey, Bill, 189
Bank Compartment, 73
Barberton, 20, 21
Barclays National Nominees,
  64
Barlow Rand, 49
Barnato, Barney, 17, *17*
Basal Reef, 48
Basutoland, 30
Bath and Portland Group, 205,
  210
Batts Combe, 216, *251*
Beatty, Chester, 139
Bechuanaland, 30
Beckett, Michael, 232
Beit, Alfred, 21, 27, 28, 32, 37,
  48
Bennet, Melvyn, 222
Big Boy Fault, 73
Big Syncline, 98
Bishop, Jim, 133
Blaawbank, 42
Black Mountain mine, 98–100,
  100–2, *100*, *102*, *103*, 110
Blainey, Geoffrey, 160, 171
Blue Circle Aggregates, 205
Blue Tee Corp., *117*
Blyvooruitzicht, 46, 48, 71, 75
Boberschmidt, Richard (Dick),
  127, 128–9, 131
Boer War, 35, 37, 39
Boers, 31
Bolivia, 175
Botswana, 94
Brazil, 175
Brisbane, 165
British Iron and Steel Research
  Associations, 83
British Quarrying Company,
  203
British Rail, 210, 215, 224
British South Africa Company,
  32
Broken Hill, 98
Broken Hill Proprietary, 164
Brooks, Sam, 246
Brown, Colin, 190
Brown, William K. (Bill),
  109–10, *109*, 114, 116, 118,
  124
Brown-Strauss, 106
Bulolo, 161–3, *161*, *162*

Bulolo Gold Dredging Ltd.,
  163
Burnie, 165
Butler, John, 175

California, 14, 110–11
Canada, 110
Canberra, 165
Cape Colony, 30–1
Cape Province, 98
Capel, 189, 196
Carbon Leader, 46, 48, 66, 70
Carbon Leader Reef, 42
Carlin, 110, 143, 153–4, *153*,
  154, *155*
Carlin Gold Mining Company,
  153
Carleton Jones, Guy, 41, *41*,
  42, 44, 45, 91
Cartwright, A.P., 73, 75
Cassidy, Peter, 192
Cement Products Corporation
  (CPC), 125, 126, 132–4
Cementation, 45, 75
Central Mining, 46, 48, 49
Central Rand, 41, 46, 56
Cerrillos, 110
Chamber of Mines (South
  Africa), 89 *
  Mines Safety Division, 83
Chambers, J.A., 29
Chartered Company, 50
Chemgold, 114
Chile, 110, 149, 150, 248
Chimney Creek, Nevada, 118,
  246
Chipping Sodbury, *200*, 202–3,
  202, 204, 209, 224–5
Chocolate Mountains, 116–17
Christopherson, Douglas, 44
Clay, Henry, 30
Clearwater, 132
Cleaver, Douglas, 204–5
Clydesdale, 96, 97
Coalbrook, 96
Coln Gravel, *212*, 213
Colombo, 220
Columbia School of Mines,
  138
Consolidated Gold Fields
  Australia, 51
Consolidated Main Reef, 70
Continental Oil (Conoco),
  121, 156
Conzinc Riotinto Australia
  (CRA), 165, 167
Corporate Finance, 102
Craig Yr Hesg, *200*
Cross, Harry, 76–8
Cudgen South, *195*
Cyprus Mines, 181

Davies, H.E.M., 27, 28, 36,
  228
Davis, Joe, 106, 108, 109

Dawes, Wally, 190
De Beers Consolidated Mines,
  17, 48, 167
Deason, John, 160
Deelkraal, 56, 59, 63, 68, 85
Delagoa Bay, 27
Devine, Joe, 132, 133, 134
Dick-Lauder, J.E., 20
Doornfontein, 71, 80, *81*, *82*
Dopper Kerk, 32
Driefontein Consolidated, 47
Du Boulay, Louise, 238–9, 245
Duminy, F. J., 87
Duncan, Val, 144, 249
Durban, 76

East Driefontein, 47, 56, 62–3,
  64, 81, 85
East Midlands airport, 220
East Rand, 27, 38, 42, 56
East Rand Proprietary, 68
Edison, Thomas, 112
Ellis, Martin, 211
Empire mine, California, 141
Eneabba, 167, 189, 191–2
Enid, 120
Enterprise Goldmines, 164
Erdeljac, Dan, 128, 129, 131–2
Erroll of Hale, Lord, 244
Escom, 96
Exxon, 172

Failing, George E., 106, 108
Failing, George E. Company,
  120–3, *122*
Falkland Islands, 220
Far West Rand, 48
Far West Rand Dolomitic
  Water Association, 71
Farren, Chris, 222
Federal Trade Commission
  (US), 145–6
Fells, Peter, 64, 92, 164, 238,
  240
Fenton, Colin, 56, 59, 92, 247
Filippone, Vincent, 52, 106,
  108
Finucane Island, 187
Fleischcer, S.R., 63
Flin Flon mine, 144
Florida, 125, 126
François, Albert, 75
François Cementation, 45, 75

Game Conservancy, 217
Garstin, N., 20
Gatwick, 220
Geissler, Hilmer, 165–8
Gemsbokfontein, 42
General Mining, 28, 45, 49
General Mining-Union
  Corporation (Gencor), 49,
  96, 244
Georgia, Alabama, 126
Gibbs, Stan, 63
Glenover Phosphate, 50

Glynn, Christopher, 125, 225, 226
Gnodde, Drury, 96, 97, 98
Goerz, Ad & Co., 28
Gold Fields American Corporation, 52, 104–35
Gold Fields American Development Company, 38
Gold Fields Australian Development Company, 158–99
Gold Fields Coal, 97
Gold Fields Deep Ltd., 37
Gold Fields Environment Trust, 243
Gold Fields Mining Corporation, 109, 194
Gold Fields of South Africa Ltd., 21–51, 54–93
Gold Mines of Australia Ltd., 158
Golden Quarry, 20
Goldsworthy Mining Ltd., 181–7, 182, 183, 184
Goldsworthy, Mount, 167, 181–7, 182, 183, 184
Gould, Charles, 169
Granduc, 144
Great Linford, 217
Greenside Colliery, 96
Greenwoods, 202
Griqualand, 17, 30
Guinea Airways, 163
Guise, George, 159

H.E. Proprietary, 52
Hallblauer, Dieter, 42
Hamersley, 182
Hammond, John Hays, 32, 35, 37–8, 41, 104, 158
Hannaford, Bob, 181
Harris, Lord, 36, 228
Harvie-Watt, Sir George, 50, 51, 230
Hatch, F.H., 29
Heathrow, 220
Hichens, Antony, 54, 127, 145, 146, 156, 168, 205, 232, 235, 244, 248
Hickson, Robin, 111, 112, 113, 116–7
Homestake, 110, 117, 154
Hoover, Herbert, 41, 139
Hope, Robin, 50
Hornblow, Mike, 215
House of Eckstein, 28
Houston, 127, 128
Hudson Bay Mining and Smelting Company, 144
Hutsell, John, 127
Hydro Conduit, 125, 126, 127, 128–32, 130, 133, 226, 243

International Loss Control Institute, Atlanta, 83
International Mine Safety Rating Programme, 83
International Monetary Fund, (IMF), 150, 245
International Tin Council, 175
Irvine, 127

Jack, John, 37
Jacksonville, 194
Jacobs, Laurie, 169
Jameson, Leander Starr, 32, 32, 35

Jamestown, 39
JCI (Johnnies), 45, 49
Jimmy the Pieman, 169
Johannesburg, 23, 28, 29, 32, 36, 37, 37, 61, 78, 96; Fox Street, 55, 103
Johannesburg Consolidated Investments Corporation, 28
Johndrow, Art, 120–3, 108–9
Johnnies, (JCI), 45, 49

Kainantu, 163
Kalahari Desert, 98, 100
Kalgoorlie, 51, 158, 160
Kansas City, 131
Kennecott, 152
Kennedy, J.F., 35
Kidd, Alex, 191
Kimberley, 10–11, 15–17, 21, 25, 28, 30, 167
Kinver, Peter, 100, 101
Klerksdorp, 48
Klerksdorp Mines, 48
Kloof, 56, 61, 62–3, 66, 76, 78, 79, 79, 88
Knoxville, 117, 118 119
Knoxville Iron Company, 106, 117, 119–20
Knoxville Iron and Steel Service, 119, 120
Kober, Alfred, 184
Kraft, Phil, 141
Krahmann, Rudolf, 41, 42–4, 43, 45
Krones, Bob, 106, 145
Kruger, Paul, 32
Krumb, Henry, 137
Kuruman, 100
Kvansnicka, Jim, 123
Kyle, 126

Lae, 161, 165
Lake View and Star, 40, 158, 160
Lakewood, Colorado, 118
Lamont, Thomas, W., 137
Lancer, S., 83
Las Vegas, 126
Lead, South Dakota, 154
Lechlade, 212, 213, 222
Leeudorn, 237
Lena Goldfields, 143
Leslie Williams Memorial Hospital, 86–7, 86
Lesotho, 94
Libanon, 42, 46, 56, 61, 62, 71, 73
Lightcap, Bill, 127–8
Livermore, John, 154
Livingstone, David, 100
Lloyd-Jacob, David, 52, 145
Lobengula, 31–2
Lombard, Herman, 63, 86
London, role of headquarters, 228–50; Charles II St., 229, 247; Moorgate, 54, 228, 247; Old Jewry, 228
Louw, Adriaan, 63, 91
Lydd, 212
Lyell, Mount, 168–75, 170, 173, 174–6
Lynch, Cornelius, 169
Lynch, Michael, 160, 196
Lyons, Joseph, 180

McCall, Donald, 50, 159, 202

McKinsey, 125
Maclaren, Malcolm, 44
McLaughlin Mine, 117
MacLeod, H., 20
McMahan, John, 123
McNab, A.J., 141, 141
McRae, Stuart, 217
Madigan, Sir Russel, 165
Magma mine, Arizona, 139, 139, 140, 141, 141, 143, 144–5, 148
Main Reef, 27, 38, 42, 46
Malawi, 94
Malozemoff, Alexander, 141
Malozemoff, Plato, 137, 137, 141–3, 144–5, 146, 148–9, 150–1, 156
Manila, 165
Marasbad, 20
Marley, 226
Mashonas, 32
Matabeleland, 31, 32
Matla, 96
Maybury, Sir Henry, 203, 203, 220
Mennell, C.S., 49
Merensky, Hans, 41, 44
Mesa, Arizona, 130
Mesquite mine, 114, 115, 116–17
Metalion Ltd., 52
Metallgesellschaft, 164
Miller, Derek, 184, 186
Mines Safety Division, Chamber of Mines, 83
Minorco, 243
Mitchell, John, 176–9, 180
Moffat, John, 31
Moffat, Mary, 100
Moffat, Robert, 100
Mohawk, 131
Moligal, 160
Montgomery Ward, 132
Moreton Island, 195
Morgan, Hugh, 150
Morgan, J.P., 139, 141
Morgan, J.P. & Co., 137
Morrow, Bill, 120
Mortimer, Gerald, 50, 159, 181, 202, 203, 216
Motapa, 78
Mount Fubilan, 164
Mount Goldsworthy, 51, 167, 181–7, 182–4
Mount Lyell Mining and Railway Company, 51, 159, 168–75, 170, 173, 174–6
Mount Pleasant airfield, 220
Mozambique, 89, 92, 94
Munro, Roy, 145, 237, 240
Murray, Brian, 186
Mustar, 161

Napa Valley, 110
Naples, 133, 134
Natal, 37
National Crushed Stone Association, 134
Ndebele, 31
Nesbitt, R.A., 20
Neumann, S., 37
Nevada, 111, 125
New Clydesdale, 96
New Consolidated Gold Fields, 40
New Guinea Goldfields, 163
New Union Goldfields, 50

New York, Helsley Building, 104, 104
New York Central Railroad, 104
Newconex Holdings, 51
Newman mine, 182, 187
Newmont Mining Corporation, 14, 48, 52–3, 98, 98, 110, 135, 137–57, 168, 234, 239
Newmont Oil Company, 143
Newport Beach, 125–8
Nigel Mine, 38
Nigeria, 220
Nixon, Richard, 56, 63
NKK, 187
Noife, Alberto, 74
Nolan, Sidney, 196
North Driefontein, 56
North Star mine, 141
Northam Platinum, 247

Oats, Richard, 160
Oberholzer Compartment, 71
Offham, 215
OK Tedi, 164
Olifants River, 20
O'okiep Copper Company Ltd., 97, 98, 98, 141, 144, 147
Oppenheimer, Ernest, 29, 48, 49, 50, 139, 249
Oppenheimer, Harry, 155, 244, 249
Orange Free State, 47, 48, 70, 71, 76
Orrell-Jones, Keith, 126–7, 127, 132, 226
Ortiz, 110, 111–14, 111, 112, 113, 116
Ortiz, José, Francisco, 111

Palabora mine, 144, 144, 146, 147–8
Papua New Guinea, 158, 160–4, 165
Parabuda mine, 182
Parker, Gordon, 157, 235
Patterson, Dick, 163, 165, 171, 173
Peabody Holding Company Incorporated, 143, 143, 145, 151–3
Peak Hill, New South Wales, 164
Perth, 165
Peru, 110, 150
Phelps Dodge Corporation, 98, 110
Philippines, 150, 165
Phillips, Lionel, 32, 35
Picacho, 114
Pilgrim's Rest, 20
Pine Creek, 164, 197
Pinter, Joe, 188
Pitts, Jim, 119
Placer Development, 161
Plumbridge, Robin, 55–6, 91, 103, 147, 234, 240, 243
Porgera, 249
Port Hedland, 185, 186–7
Portland, Oregon, 132
Potier, Gillie, 50
President Brand, 48
President Steyn, 48, 68
Pretoria, 20, 24, 33, 71
Prevention of Accidents Committee, 83
Pullinger Brothers, 42, 44, 45

Racheff, Ivan, 119
Ramsay, Robert H., 147
Rand Mines, 28, 32, 48
Rand Mutual Hospital, 86
Rand Refinery, 70
Rand Selection Corporation, 49
Randfontein, 44
Randfontein, Estates, 70
Raynor, Doug, 215
Reagan, Ronald, 193
Red Cross, 83
Redland, 226
Reinecke, Leopold, 41, *41*, 45
Renafontein, 42
Renison Ltd., 159
Renison Goldfields Consolidated Ltd. (RGC), 159, 163, 164–8, 172, 175, 181, 188–99, 235, 243
Renison Tin, 168, 175–9, *176*, *177*, *178*, 180, *181*
Rensburg, Peter van, 62, 71
Resurrection mine, 143
Rhodes, Cecil John, *12*, 13–23, *14*, 20, 26–8, 30–2, 35–6, 41, 48, 228, 249
Rhodes, Frank, 20, 32, 35
Rhodes, Herbert, 15, 20
Rhodesian Anglo-American, 139
Rhodesian Selection Trust, 139
Richard's Bay, 97, 190, 192
Rio Tinto Zinc, 144, 148, 156, 237
RMC, 133
Roberts, Max, 159, 164, 172, 199
Roberts, Ralph, J., 154
Robinson Deep mine, 37, 38, 41, 62
Robinson Gold Mining Company, 27
Robinson, J.B., 21
Roe, Peter, 107, 247
Rooiberg, 97, 98, 102
Rooiberg Minerals, 50
Roosevelt, F.D., 244
Rooyen, Bernard R. van, 102, 110
Rudd, Charles, 15, *15*, 21, 26, 28, 31, 35–6, 228, 249
Rudd, Henry, 15
Rudd, Thomas, 228
Rustenburg Platinum Mines Ltd., 40, 43

Ryan Consolidated, 242

St Ives, 222, 223
St Petersburg, Florida, 126
Salt Lake City, 139
San Manuel mine, 139, 145
Sandel, Luiz, 74
Sagaram, Gerry, 187
SASOL, 96, 103, 243
Saunt, Gatty, 203, *203*
Scallan, Dick, 192
Schroders, 159
Searles Lake, 40
Searls, Fred, 138, 141
Seattle, *124*
Secrist, Richard, 118, 126
Segal Sidney, 159
Shakesby, Roger, 163, 165
Shay Gap, 181, 184–7
Sheba Mine, 20
Shepherd, Michael, 194
Sherritt, Gordon, 144, 156
Shetlands, 220
Siberian Lena Goldfield, 40
Sierra Leone, 192
Silver Queen mine, *139*
Silverstone, 220
Simmer, August, 37
Simmer and Jack mine, 37, 38, 62, 83
Sishen, 100
Skelton, Len, 189–90, 194
Skytop Brewster, 52, 107–9, *107*
Smith, Henry DeWitt, 141
Somerford Keynes, *210*, 211
Sotho, 18
South African Gold Trust and Agency Company, 28
South African government, 56
South West Africa Company, 50
Southampton, 213, *214*
Southern Peru Copper Corporation, 144
Southern Transvaal, 71
Soviet Union, 92, 94
Spaarwater, 41
Spence, Charles, *127*, 205–6, 226–7, 234
Springbok, 98
Stanley, Sir Henry, 13
Stanton Harcourt, 222
Stanwick, *205*
Star Diamonds, 50
Steel Service Company, 106

Sticht, Robert Carl, 169–71
Strathmore Company, 48
Struben, Harry, 20
Strubens, 24
Sub Nigel Mine, 38, 41, 44, 62, 63
Sumburgh, 220
Sunrise Hill, 184, 186
Sutton Courtenay, 215, 217, 220
Swaziland, 94
Sydney, *198*, 199
Sykes, Allen, 232

Tarbutt, Percy, 28, 41
Tarmac, 132, 133, 226
Tasman, Abel, 169
Tasmania, 168–80
Taute, Dr, 75, 79
Telfer, *152*
Texas Gulf Corporation, 156
Theron, W.C., 74
Thompson, Guy, 133
Thompson, William Boyce, 48, 137–41, *137*
Tom Price Mine, 182
Transvaal, 31, 32–5, 70
Transvaal Coal Owners Association (TCOA), 96–7
Trollope, Anthony, 21
Trollope, Mr., 87
Trudeau, 110
Truter, Albert, 23, 45, 46
Tsumeb mine, 144, 147, *148*
Twidle, Tim, 101

Uitval, 42
Unimet Corp, 106
Union Corporation 28, 45, 49
Union Tin, 50, 98
United Nations, 94
United States, 104–57
Upington, 100

Vaal, River, 47
Vanderbilts, 104
Vascoe, A., 74
VCR, 66, 70, 73
Venda, 20
Venterspost, 42, 46–7, 56, 58, 60, 71, 73
Ventersdorp Contact Reef, 42, 46
Vlakfontein, 41
Vogelstruisbult, 41
Vryburg, 100

Wallingford, 217
Wanderer Mine, 78
Waratah, 179–80
Warren & Wetmore, 104, *104*
Waterboer, 17
Wau, 161, 163, *178*
Weinstock, Lord, 128
Welkom, 48
Wernher, 37
West Driefontein, 43, 46, 56, 57, 63–73, *65*, *66*, *67*, 73, 85, 87
West Rand, 42, 56
West Witwatersrand Areas Ltd., 44, 54
West Wits Area Company, 46
West Wits Line, 41, 44–9, 56, 62, 93, 103, 237
Western Deep Levels, 49, 68–9, 70, 71, 75
Western Mining, 150, 168
Western Ultra Deep Levels, 48
Westinghouse Air Brake Company, 120
Westminster Gravels, 205
Whatley, 207–10, *209*
Whillier, Austin, 76
Wilson, Ned, 189
Wiluna Gold Corporation, 40
Wiluna Mine, 158
Witwatersrand, 20–1, 23, 42–4, 71, 76
Witwatersrand Chamber of Mines, 29–30
Witwatersrand Native Recruiting Corporation Ltd., 29
WMK, 125, 126
Wood, Humphrey, 109, 205, 232
World Bank, 150
Wright, Isabel, 224, 225

Yelland, David, 207
Yilgara Block, 164
Yola, 20
Yuba, 38

Zaire, 149, 150
Zambia, 149, 150
Zeehan, 179
Zimbabwe, 78
Zincor, 98, *101*
Zoutpansberg, 20